Earth's Changing Climate

Professor Richard Wolfson

THE TEACHING COMPANY ®

PUBLISHED BY:

THE TEACHING COMPANY
4151 Lafayette Center Drive, Suite 100
Chantilly, Virginia 20151-1232
1-800-TEACH-12
Fax—703-378-3819
www.teach12.com

ISBN 1-59803-296-8

Richard Wolfson, Ph.D.

Professor of Physics and Environmental Studies, Middlebury College

Richard Wolfson is the Benjamin F. Wissler Professor of Physics at Middlebury College. He did his undergraduate work at MIT and Swarthmore College, graduating from Swarthmore with a double major in physics and philosophy. He holds a master's degree in environmental studies from the University of Michigan and a Ph.D. in physics from Dartmouth. Dr. Wolfson's published work encompasses such diverse fields as medical physics, plasma physics, solar energy engineering, electronic circuit design, observational astronomy, theoretical astrophysics, nuclear issues, and climate change. His current research involves the eruptive behavior of the Sun's outer atmosphere, or corona, as well as terrestrial climate change and the Sun-Earth connection.

Professor Wolfson is particularly interested in making science relevant to nonscientists and to students from all walks of academic life. His textbook, *Physics for Scientists and Engineers* (Addison Wesley, 1999), saw three editions and has been translated into several languages; it was followed by *Essential University Physics* (Addison Wesley, 2006). Dr. Wolfson's books *Nuclear Choices: A Citizen's Guide to Nuclear Technology* (MIT Press, 1993) and *Simply Einstein: Relativity Demystified* (W.W. Norton, 2003) exemplify his interest in making science understandable to nonscientists. His latest book, *Energy, Environment, and Climate* (W.W. Norton, 2007), is an undergraduate textbook accessible to general readers. He has also published in *Scientific American* and writes for *World Book Encyclopedia*. His previous courses for The Teaching Company include *Einstein's Relativity and the Quantum Revolution: Modern Physics for Nonscientists* (1999) and *Physics in Your Life* (2004). Dr. Wolfson has spent sabbaticals at the National Center for Atmospheric Research in Boulder, Colorado; at St. Andrews University in Scotland; and at Stanford University.

Table of Contents
Earth's Changing Climate

Earth's Changing Climate

Scope:

Is Earth warming? Yes: Recent years have been the warmest our planet has experienced in the past millennium and quite possibly for 100,000 years or more. Are we humans responsible? At least in part: We've increased the concentration of heat-trapping gases in Earth's atmosphere to levels not seen in the past million years and probably much longer. What does the future hold? A continued warming trend, increased melting of polar and glacial ice, rising sea level, altered weather patterns, and more intense precipitation and storms. What will be the impact? That's less certain—but it will surely include changes in agriculture, flooding of low-lying areas, extinction of species, and economic consequences. Sudden, large-scale disruptions of Earth's climate are less likely but lurk as increasing possibilities beyond the current century.

How do we know that Earth's climate is changing? How can we project future climate change? Why do we attribute climate change to human activities? This course explores the solid science behind climate change, the myriad evidence pointing to ongoing global warming, and the modeling techniques used to project future climate. The course assumes no background in science and is designed to give students a firm understanding of the scientific issues that underlie climate change—issues that must ultimately guide public policy in the coming decades.

This is a course about the science of climate change. It is not an advocacy program. Other groups and individuals, through writings, films, and public information campaigns, attempt to educate and, simultaneously, to influence public behavior. This course seeks only to educate. As such, it presents evidence for what has become an overwhelming scientific consensus on global climate change and its causes. You should come away from the course better equipped to make your own judgments about the appropriate public policies to cope with climate change, but this course won't make those judgments for you. That said, the course at its end acknowledges the need for action on climate change—but it suggests a range of alternatives rather than dictating one course of action.

The course is presented in 12 lectures. The first half explores our scientific understanding of climate, while the second half deals with future climates and the human role in climate change. The first lecture, "Is Earth Warming?" gives an introduction to the entire course, distinguishes clearly between science and policy issues, and presents evidence for recent planetary warming in the context of the 150-year record of measured global temperatures. The second lecture, "Butterflies, Glaciers, and Hurricanes," examines a host of phenomena—from species migrations to ice melt to storm intensities—that, along with rising temperature, point to ongoing climate change. Lecture Three, "Ice Ages and Beyond," looks at ancient climates, inferred from a wide range of physical, chemical, and biological indicators. Lectures Four and Five, "In the Greenhouse" and "A Tale of Three Planets," explore the scientific principles that establish climate and show how the climates of our neighbor planets Venus and Mars confirm our understanding of Earth's climate. These lectures also include a brief discussion of the nature of scientific theories and of certainty and uncertainty in science. The first half of the course ends with "Global Recycling," a look at the complex cycling of materials—especially carbon—in the atmosphere, oceans, soils, and biological systems.

Lecture Seven, "The Human Factor," confronts the evidence that the climate change of recent decades is predominantly the result of human activities—largely but not exclusively the combustion of fossil fuels. Lecture Eight, "Computing the Future," shows how computer models combine basic scientific principles with complex feedback effects and observations of past climates to project scenarios for future climate change; the lecture also presents test results that give us confidence in computer climate models. The ninth lecture, "Impacts of Climate Change," explores expected changes in weather patterns, heat waves, sea level, ice sheets, drought, and other effects of a changing climate. Lecture Ten, "Energy and Climate," takes a realistic look at the relatively few basic alternatives we have to climate-altering fossil energy sources. "Energy—Resources and Alternatives" follows with a more detailed look at the prospects for alternative energy technologies and their role in mitigating climate change. The course ends with Lecture Twelve, "Sustainable Futures?"

Lecture One
Is Earth Warming?

Scope:

Earth's climate is changing. Recent decades have seen a steep rise in global temperatures. Thermometer-based temperature records show a pattern of temperature variation that spans 150 years, from the mid-19th century to the present. What does this pattern tell us? How do we take our planet's temperature? This lecture explores these scientific questions and shows how climate scientists arrive at values for global temperatures.

Questions about climate change and its causes are squarely in the realm of science. Questions of what to do about climate change are in the realm of public policy. With climate change, issues of science, policy, and political opinion often become muddled. This course is about science, and what's presented here is based on solid scientific evidence developed through the course of research, peer-reviewed publication, and the emergence of scientific consensus. Science can help guide policy, but science alone can't dictate the best policy decisions. This first lecture includes an introduction to the course that clearly distinguishes the realms of science and policy.

Outline

I. This course deals with *climate*, not *weather*. *Climate* refers to long-term trends; *weather*, to short-term variations in atmospheric conditions. We jump right in with a look at Earth's average temperature over the past 150 years.

    A. There will be lots of graphs in this course! My obligation is to explain clearly what each graph shows, what the axes are, what units are used, where the data come from, and what the uncertainties are.

    B. We begin by looking at global average temperatures since 1860, expressed as deviations from the 1961–1990 average temperature. See Figure 1.

        1. Temperatures are in degrees Celsius, used almost universally in science; 1 degree Celsius is 1.8 degrees Fahrenheit.

2. It is easier to measure temperature changes accurately than to measure temperature itself; thus, climate trends are often shown as deviations.

3. Data are from thermometers; more on this shortly.

4. Temperature deviations in the early decades are good to about ±0.2°C; by 1950, this uncertainty in the global temperature record drops to roughly ±0.05°C.

C. The 150-year temperature record shows a lot of variation but can be divided into several major sections:

1. A roughly constant average temperature in the late 19th century.

2. A rise in the early 20th century.

3. A leveling off or slight decline in the mid-20th century.

4. A steep rise in the final decades of the 20th century, continuing into the 21st century.

D. What causes this pattern of temperature variation, and how typical is it for our planet? That's a major theme of this course, but here's a quick summary:

1. Variations in the early part of the 150-year temperature record are largely explainable by natural variations within the climate system, by volcanic eruptions, and by variations in the Sun's energy output.

2. Variations in recent decades are largely the result of human activities, predominantly but not exclusively the burning of fossil fuels. The second half of the course will detail the evidence for a human impact on Earth's climate.

E. So far, we've discussed only the global average temperature. Regional temperatures show considerably greater variability, and land temperatures have risen more than ocean temperatures. The temperature increase has also been greater at high latitudes. (Take another look at Figure 1, which also shows Northern Hemisphere land temperatures.)

F. The bottom line is that Earth's average temperature has risen about 0.65°C since the start of the 20th century, with nearly all regions of the globe experiencing a temperature increase. This may not seem like much, but we'll see in subsequent lectures why even such a small rise in the global average temperature is climatologically significant.

II. How do we take Earth's temperature?

  A. A single thermometer won't do! But since the mid-19th century, there have been enough data available from thermometer-based measurements to compute an average global temperature at Earth's surface. Data sources include:

  1. Air temperatures from land-based weather stations (surface air temperature, or SAT), with thermometers usually placed 1–2 meters above the surface.

  2. Marine air temperature (MAT), taken just above the sea surface from ships and buoys.

  3. Sea-surface water temperature (SST), taken from ships.

  B. All these measurements require corrections to ensure accurate global averages:

  1. Corrections for instrumentation and placement.

  2. Corrections for water-temperature sampling.

  3. Corrections for the *urban heat island effect*.

III. Now let's turn to a discussion of science, policy, and this course.

  A. Science deals with the facts and governing principles of physical reality. Science is never 100% certain, but neither is anything else. Science develops through observation, experimentation, theorizing, and publication in open, peer-reviewed literature. Scientific consensus emerges when the results of observation and experiments overwhelmingly support a given idea or hypothesis. Science can be quite certain of "big picture" ideas without being certain of every detail. Such is the case with the state of climate science today.

  B. Public policy involves society's decisions about what to do in the face of many factors—including the findings of science. Science can guide policy by providing factual information and projections of future conditions, but science alone can't determine the right policy decisions.

  C. Ongoing changes in Earth's climate have significant implications for public policy. This situation has led to a muddling of science and policy questions in the minds of the public and of many policymakers. This course is about climate science, not politics or policy debates.

  D. The two halves of this course deal with:

1. Our basic scientific understanding of Earth's climate system, including a look at past climates and the principles that determine climate.
2. Our evidence for a human impact on climate, our understanding of the causes of that impact, and the implications for future climate and for human use of energy.

Suggested Reading:

Houghton, chapter 1.

Wolfson, chapter 14, section 14.1 (skip "Going Further Back").

Going Deeper:

Harvey, chapter 5.

Intergovernmental Panel on Climate Change (IPCC) 4, chapter 3, section 2.

Web Sites to Visit:

Climatic Research Unit, University of East Anglia, http://www.cru.uea.ac.uk/.
Probably the most authoritative source for the instrumental temperature record discussed in this lecture.

Questions to Consider:

1. Can you tell if the 150-year global temperature record discussed in this lecture represents anything unusual in the history of Earth's climate? If not, what additional information would you need to decide whether or not recent climate change is unusual?

2. Describe two of the difficulties in arriving at an accurate value for the global average temperature and tell how scientists correct for these difficulties.

Lecture One—Transcript
Is Earth Warming?

Welcome to *Earth's Changing Climate*, a course with The Teaching Company. If you're new to my courses, welcome. If you've viewed or listened to my courses, *Modern Physics for Non-Scientists* or *Physics in Your Life*, welcome back. This is a rather different course; we're looking at a much more contemporary scientific issue, and we'll plunge right in with the first lecture. The question it asks is: Is Earth warming? I hope in the course of this first lecture to answer that question, at least in the context of the last 150 years. Then in subsequent lectures, we'll go on further.

It happens to be a beautiful January day here in the Washington, DC area where The Teaching Company is located. It's sunny; it's climbing through the 40s and into the 50s—just a nice day. Now that's a statement about the weather. I hope the weather is nice where you are, but having seen the weather reports this morning, I know there are parts of the United States that are digging out from blizzards. Weather varies, and weather varies a lot from day to day.

I want to make a very clear distinction between weather and climate. This is a course about climate. It's not a course about the weather; it's not about predicting what tomorrow's weather will do. It's about long-term trends in the weather; how the weather changes over years and decades, not what happens from day to day. Having said that it's a beautiful day here in Washington, I'll also point out that it is running, nationwide today, some 20 degrees warmer than average for early January. I'll also point out that the last few months in my home state of Vermont have been abnormally warm, remarkably warm. We've had only one significant snowfall. We haven't seen any temperatures below zero, or even in the single numbers. Those are statements about climate, and they're statements about climate because they're comparing what's happening today—and not just today but over say the last few weeks or months—with what has happened in the past over similar time intervals.

Climate is about the long-term trends in the weather, the long-term trends in what's happening in the atmosphere, what's happening to temperature, what's happening to precipitation, and so on. Weather refers to the short-term, local, immediate events in the atmosphere. I want to emphasize that distinction. This is a course about climate; is climate changing? Particularly, we're asking the question, here today in this first lecture: Is Earth warming? Let's plunge right in with a graph that looks at the Earth's average temperature, the global average temperature, some kind of representative temperature of the entire planet over roughly the last 150 years. I'm going to pause here and give you a little aside on graphs. You know the statement that a picture is worth 1,000 words? I think of a graph as being worth 1,000 numbers. I'm going to show you lots and lots and lots of graphs in this course because there are lots and lots of numbers that summarize the state of climate. The easiest way to see what those numbers are doing is to look at them graphically.

I have some obligations to you when I show you a graph. By the way, those of you who are listening to this course in your cars or otherwise listening on audio, you have access to these graphs in your booklet, and I urge you to look at them at a time when it's safe for you to do so—not when you're cruising down the freeway. I will try to describe as much as I can what these graphs show, so if you're in the audio audience, you'll get a sense of what's going on here. Not only are the graphs by the way in your booklet, but there's also a list of the sources of all the data that went into these graphs, so that you can understand where they've come from. My obligation is to tell you what the graph is showing, to tell you what the two axises—the horizontal and vertical axis—of the graph show you, and to tell you something about how uncertain we are about the information in the graph because science is not 100 % certain about anything. Before you say well, wait a minute, what about science? Science isn't very useful, then. Nothing else in life is certain, and science can be quite certain. We can put numerical certainties and uncertainties on our information, and I will try to tell you when we look at graphs roughly how certain or uncertain we are.

Let's go back to this graph of the average global temperature over a time span since 1860, roughly. The first thing to notice is that the vertical axis here is in degrees Celsius (°C). In the United States, we tend to use degrees Fahrenheit (°F) for temperature measurement; the rest of the world uses Celsius. Degrees Celsius is exactly 1.8 degrees Fahrenheit. So if you want to make a very rough conversion, if you hear a temperature in Celsius, or a temperature change in Celsius, you can roughly double it to get the temperature in Fahrenheit. There's also that 32-degree offset because water freezes at 32°F, but at 0°C, so there's a little bit of complication in converting those. But

I'll mostly be talking here in this course about temperature changes, and there, a temperature change of 1°C is equal to a change of 1.8°F, or approximately 2°F.

It's easier to measure deviations in temperature than to pin down the actual exact global temperature. So, this graph—and a number of the other graphs I'm going to show you—show not temperatures, but deviations in temperature from some average value. In this particular case, the average value that the climatologists who prepare these graphs use is from the 1961–1990 average temperature. It doesn't matter what average we take, but once we take one, we're looking at deviations from that average. The particular graph I'm showing you now shows deviations from that average. I've done an area graph where I've filled in the area between the average temperature line and the line representing the actual temperature. Consequently, you'll see a number of areas where the graph is below that line, and the graph is extending downward. Those are times when the temperature was below the 1960–1990 average. You'll see regions where the temperature is clearly above that average, and you'll see some fluctuations where it's going back and forth either way.

Why do we do these deviations rather than to try to get a global temperature? As I'll describe later, these measurements come from thermometers all around the globe; different kinds of thermometers, different places, some in the ocean, some in land, some on mountains. It's much easier to talk about what changes are happening at a given station and incorporate those into a global average change in temperature, than it is to talk about the actual temperature calculated from the temperatures at all these different spots. So we'll see very frequently temperatures expressed as deviations from some average. As I mentioned, these data come from thermometers, actual physical thermometers, like you would use to measure temperatures. I'll say a little bit more about how that's done later.

I've mentioned that I should talk about uncertainties in our information about these quantitative data. In the early part of this record, in the late 19th century and into the early 20th century, these temperatures are accurate to probably about 0.2°C. That is, if you look at the graph and you say "Well the temperature was down by 0.5°," it might have been down by only 0.3°, or it might have been down by 0.7°. In other words, ±0.2° is roughly our uncertainty. By the middle of the 20th century and beyond, we're pretty sure that these temperatures are correct to within about 0.05°C, about 5/100 of a degree Celcius. The reason for that improvement is an increase in the number of stations we have reporting temperatures from around the globe. Having understood this graph, let's take a look at what it shows. There's a lot of variation, first of all. The temperature fluctuates from year to year substantially. But there are several major sections in this graph, and they're worth looking at.

First of all, from about 1860 into roughly the year 1900, the graph is roughly flat. I say very roughly because there's a lot of natural variability in the climate, and you can see the temperatures fluctuating quite a bit. But overall, if you kind of drew a line representing the average trend here, the temperature has stayed roughly constant from about the mid-19th century, when this record starts, to about the beginning of the 20th century. There's a rise in the early 20th century, quite a pronounced, substantial rise in temperature. Then there's a leveling off in the mid-20th century, maybe even a slight decline. Then there's a rather steep rise that characterizes the final roughly three decades of the 20th century, and it continues into the 21st century.

Our question is—for this lecture, and for in fact, the entire course—is what causes this temperature variation, and how typical is it for Earth? Is this just what Earth naturally does, or is there something unusual here? That's a major topic of the course, but I want to give you right now just a very quick summary. We think that in the early part of the 150-year record, the variations there are caused largely by natural variations of three types. First of all, the climate system itself is a very complex thing, and has internal oscillations. The El Nino phenomenon that you've probably heard of is an example of an internal oscillation in the climate system. It just happens naturally. The climate doesn't stay absolutely flat and steady. That's some of the variation.

Volcanic eruptions, of which there were a significantly large number in the early 19th century in particular, put dust into the atmosphere, and that exerts a cooling effect on climate. We think volcanism has a significant role to play in the early part of this record. There have been some significant—not substantial, but significant—variations in the power output of the Sun, and that's what, of course, supplies Earth with its energy, so that causes some of the variations. But we're quite certain that the variations in the global average temperature from roughly 1950 or so onward, maybe 1970 onward, cannot be explained without a human influence on climate; and particularly the dominant human influence being the emission of carbon dioxide from burning of fossil fuels. I'll have a lot more to say about that in the second half of the course. The second half will deal particularly with our evidence that there is a human effect on climate, and we're seeing it, and it's there in the last part of this temperature record that I showed you in this graph.

So far, I've discussed only the global average temperature, or rather the deviations in the global average temperature from some particular average; we've chosen 1960–1991. There's also considerable variability in regional temperatures, and I'd like to take a look at some of those. In particular, when you look at this global average temperature I've been showing you, that's the temperature averaged over land, over oceans, over the Northern hemisphere, over Southern hemisphere, and so on. If you look in particular regions, you'll find the temperature has changed rather more substantially than the global average. For example, over almost all land areas, the temperature has increased more than it has over the oceans. Since the oceans dominate the Earth's surface, the change in land temperature is much more significant than the change in the global average temperature.

This pattern I've showed you in the global average, with roughly constant temperatures in the late 19th century and early part of the 20th century—a rise in the earlier decades of the 20th century, a leveling off, and a steep rise—is evident in quite a few climate records. A colleague of mine and I recently published a paper in which we analyzed temperature records in the Northeast, and we found that this pattern prevails in almost all of the 73 individual temperature stations we looked at. Of those 73 stations, all of them showed something consistent with this kind of warming, with a lot more fluctuation. Only two of those 73 stations, for example, showed cooling. Let's take a look at some other examples of temperature records that aren't quite as global as this.

I'm going to change this graph to the same data—this is exactly the same data—now portrayed perhaps a little bit more the way a scientist might portray it, as simply a line representing the global average temperature. You see exactly the same data you saw before.

You see the roughly average temperature in the late 19th century and early part of the 20th century, and then the rise in the decades of the 20th century from about 1910 to maybe 1930 or '40. Then that leveling off, and then that fairly steep rise. It doesn't look as dramatic on this new graph because I've scaled the vertical axis to cover a broader range in temperature, and there's a reason I've done that, because what I want to do next is add on top of this the same kind of data, but now taken for just the Northern Hemisphere and just temperatures on land.

If we look at that, you see both the black curve, which is the one I showed you for the global average; and the gray curve, which is Northern hemisphere land temperatures, and you see several things. First of all, the same general pattern is there; roughly steady temperatures in the late 19th century, a rise in the early 20th century, a kind of leveling or slight fall in the mid-20th century, and then that steep rise continuing into the 21st century. However, you'll see the deviation at the very end. Into the 21st century here, it is considerably more substantial for the Northern hemisphere land temperatures than it is for the global average. You'll also see some other subtle differences. For example, especially in the early years of the record, the land temperatures are showing much more fluctuation than the sea temperatures, and there are several good reasons for that.

That gives you a sense that, when somebody talks about the global average temperature change, you want to look in a little more detail at what's been changing and where it's been changing. Really, to do that in more detail, let's take a look at a map, which will show the entire globe and give us a sense of what the temperature changes have been over the entire globe in recent times. This is not a graph, but a map. It's a kind of graphical representation of quantitative data, but it's placed on a map. Without going into all the details, what we're looking at here is a map of the Earth, and color-coded temperature changes measured in degrees Celsius per decade. For every 10 years, a 1°C per decade change means that, in 10 years, the temperature in that region went up to 1°C. At the bottom is a scale, and we're not going to go into the quantitative details of the scale, but in the middle, the white colors represent basically not much change; maybe ±0.1° per decade. Still significant, but it's not a huge change. In the far right, we're dealing with several degrees Celsius changes, the brightest red regions; and to the left, the blue regions represent places where the Earth has cooled.

The first obvious thing you'll notice is that most places have shown a warming. These are degrees per decade since the year 1950, so the encompass most of that region in the late 20th century and early 21st century, when we've seen the most dramatic rise in the global temperature. You can see very clearly that most of the planet has experienced a warming. You can see a few spots that have experienced cooling. They mostly tend to be in the oceans. You can see a place in the Pacific, off the western coast of North America. You can see just south of Greenland, there's been some cooling. There's a little place east of Asia where there's been some cooling. But most of the Earth's surfaces, including most of the oceans, have experienced some warming.

You'll see particularly a broad band of red across the Northern Hemisphere land area, which has experienced the most warming. That's a trend we'll see again and again, that the warming has been greatest over land, and it's been greatest in the Arctic. It's been a little less clear in the Antarctic, and I'll have more to say about that a little bit later. But the bottom line of all this is that, if you look at Earth's global temperature over the last 100 years or so, the Earth's temperature has risen about 0.65 or 0.7°C, a little over 1°F, in that time. That's what the global average has done. That has occurred there's been warming in almost every region of the globe, and the warming has been more substantial than that 0.65 or 0.7° would suggest in many regions of the planet. Now, 0.65°C or 1°F, or whatever it works out to, may not seem like a whole lot. But I want to emphasize that a small rise like that in the global average, representing things that are happening all over the globe, that can be climatologically very significant, and I will show you why very shortly in subsequent lectures. That's the bottom line, this warming of about 0.65°C.

I shouldn't just present these numbers; I should tell you how we take Earth's temperature. How do we do that? What won't do is to stick a single thermometer somewhere on the Earth and record an average temperature. That just doesn't work; a single thermometer won't do it. The reason I showed you temperature data dating back to the mid-19th century is because only at about the mid-19th century, about the year 1850, did we have enough data from all around the world to calculate a statistically meaningful average temperature. How did we do that? Well we have a number of sources. First of all, there are air temperatures, and they're coming from basically land-based weather stations, which place their thermometers in particular standard locations at the height of my eyes, about a meter and a half. They measure what's called surface air temperature, SAT, and they're measured from thermometers typically placed at weather stations. There are marine air temperatures, MATs, taken from buoys or ships. There are sea-surface temperatures taken from ships also, the temperature of the surface waters.

All those temperature measurements reported from all around the globe are combined with careful statistical analyses to compute a best estimate of the Earth's global average temperature. There's an art to this, and a science—a very subtle science. There are a number of corrections. For example, weather stations may move. They may replace their instrumentation, and that has to be corrected for. Water temperature sampling requires particular corrections. In the old days, they'd throw a bucket overboard, haul it up to the deck of a ship, and take the temperature. Whether the bucket is made of metal, or canvas, or wood affects how much cooling occurs. How big the ship is, how high above the water, affects how much the water may have cooled or warmed before the temperature is actually taken. Modern ships take the temperature at the intake water to their engines, so that doesn't require as much correction; that requires a different treatment. All these things have been handled.

One of the most significant changes that you may have heard about is what's called the urban heat island effect. If you imagine a weather station that, for example, was 100 years ago on the outskirts of a city, cities tend to be warmer than their surroundings because they've changed the land's reflectivity by paving roads. They generate a lot of energy themselves, which causes local heating. Well if a weather station 100 years ago was on the outskirts of a city, chances are now it's surrounded by city. Here's a visual that suggests how that might have happened. You see initially a village with a weather station on the outskirts of that village. Then, as the village grows, the urban area surrounds what was a rural weather station, and that causes the temperature of that station to go up. That will give you a false indication of global warming if you just took your measurements from such stations.

Fortunately, we can test for that effect. In these huge data sets, which have thousands of places reporting, we can simply take out the places that represent large urban areas, and we find that the global temperature record doesn't change much; maybe a few hundredths of a degree. Some of the data sets that look at global temperatures correct for that, and some don't, but the point is everybody recognizes it isn't a huge correction. It's not the reason we see the planet warming now. There's a slight effect there; if we want to get very accurate, we correct for it. But the urban heat island effect is not the reason why we have the global temperature rise. That's a quick look at the science behind how we take Earth's temperature today and how we've been able to take it over roughly the past 150 years.

I started this lecture with a question, is Earth warming? The answer I'm clearly giving you now is yes, Earth is warming, but I haven't said anything about whether that warming is unusual, what things look like going further back into the past, where unfortunately we don't have thermometers to take the record. I'll be saying a lot more about that in subsequent lectures. Let me pause a minute at this point in this first lecture and say a little bit about science, and about public policy, and about this course. Science really deals with facts; they're facts and their governing principles that tell us how physical reality works. It's never 100 % certain—I already gave you the uncertainties on that global climate record—but neither is everything else.

How do we develop science? We develop it first by observing the real world. We develop it by experimenting; perhaps less so in this field of climatology than some other fields, but nevertheless, we can do some experiments, and I'll describe those later. We theorize; we publish in peer review literature, which means we publish something. We write a paper. The editors of the journal we send it to send it out to several scientists who review it, who assess whether the science is correct, who make suggestions for changing it. The paper is published, if it's deemed suitable, it's published in the literature. Other scientists read it. They can try to duplicate the results, or they can show that these results are incorrect. So, science is a very self-correcting process.

After a while, when observation and theorizing come to a consistent picture of what's happening in the phenomenon under study, then a scientific consensus tends to emerge. In that consensus, science can be quite certain of big-picture ideas, the whole grand scheme, without necessarily knowing every detail. I want to emphasize that because that's certainly the state today of climate science. We certainly know the big picture. We know that Earth is warming; we know why Earth is warming. We know, as I indicated, and as I'll describe in the second half of the course, that at least some of that warming is caused by human activity. We know the big picture. We don't know every detail. Not knowing every detail doesn't negate our knowledge of the big picture. There are plenty of other areas of science where you can think of that also. One example is evolution, where we understand the big picture about how species evolve, but we don't necessarily know every detail of every mutation that caused one species gradually evolve into another. We do understand the big picture, and that's the state of climate science today. There are plenty of uncertainties. There are plenty of things we still need to understand, but we know the big picture.

There's another issue, and that's public policy. Public policy is about what the public ought to do about many factors, including scientific knowledge. How should we respond to our understanding of what Earth's climate is doing, for example? Science can certainly guide policy, but science isn't about policy. Some scientists stay meticulously away from saying anything about policy; other scientists will happily use their scientific knowledge to jump in and give advice to policymakers, political leaders, business leaders, and so on. That's sort of a personal choice for the scientists. Once a scientist steps outside the realm of science, he or she is then giving his or her opinion about policy. That's a different thing from science because science is this consensus-building, corrective, peer-reviewed procedure that arrives at a common understanding. A policy is based on many other things, including values, political judgments, and so on. So, science can guide policy, but science alone can't determine policy.

The content of this course, certainly more than any other Teaching Company course I've done—and certainly more than most courses I teach at Middlebury College—has significant implications for public policy, in a way that, for example, the Big Bang theory of the origin of the universe, at least at present, seems not to. It may have important consequences for how we think of our place in the universe and so on, but it doesn't have immediate policy implications. Climate change does because, to the extent that climate change is caused by human activities, then what we choose to do can affect the future climate. That situation has led to a kind of muddling of science and public policy; a lot of muddling in the mind of the public, a lot of muddling in the media. Some of the muddling has been intentional; some has not been intentional. It's raised a lot of questions in the minds of policymakers, and of the public, about climate change.

I don't want to get into questions of policy. That's not what this course is about. What I'm going to be dealing with is the science of climate, and my job is to explain to you why we understand what we do about Earth's climate, about its changes, and about the causes of those changes. I'm not going to get into the policy issues. I'm not going to recommend what we should do; although at the very last lecture, I'll give some suggestions about what future climates might look like, depending on certain paths we might or may not choose to take. This is not a course about policy. I'm not advocating policy. I'm sticking with the results of that peer-reviewed, consensus-building, theorizing, experiment, observing aspect, which is science.

The course comes in basically two halves. It's a 12-lecture course, and it basically comes in two distinct halves. The first half is going to deal with our basic scientific understanding of the climate system. We're going to look at past climate; we're going to look at principles that determine climate. Let me pause here a minute and give you a brief outline of the first half of the course, then. Here's the first half, part one, the first six lectures. The first lecture, we've just come near to the completion of, "Is Earth Warming?" We've raised that question, is Earth warming? and answered it basically in the affirmative. Earth is warming, has warmed over the last 100 years or so by most of a degree Celsius, by a little more than a degree Fahrenheit. We've taken care of that question.

Lecture Two: "Butterflies, Glaciers, and Hurricanes." Why? Because it isn't just rising temperatures, as indicated by thermometers that tell us what's happening to Earth's climate. So do things as diverse as the way species are behaving—butterflies are an example; the way glaciers are behaving—most, but not all, of them are melting; the way hurricanes are behaving—some of them are getting more intense. What does that have to do with climate change? Those may all be additional indications of climate change. Lecture Three: "Ice Ages and Beyond." Unfortunately, this thermometer temperature record can only take us back to the mid-19th century, and if we want to know whether what's happening to Earth's climate now is abnormal, or maybe just some natural fluctuation that will stop in a few decades, we need to know what Earth's climate has done in the more distant past. We'll go way back. We'll go back into the Ice Ages. We'll go back first 1,000 years or so; then we'll go back into the Ice Ages, hundreds of thousands of years. Eventually, we'll take a quick look all the way back, three billion or so years, to the beginnings of Earth's history. We'll be looking at climates in the past in Lecture Three.

Lecture Four: "In the Greenhouse." You've heard of the greenhouse effect. The greenhouse effect is crucially important for understanding why Earth's climate is the way it is; not only naturally, but if there are also human effects on it. We'll look at the science of the greenhouse effect. In Lecture Five, "A Tale of Three Planets," we'll look at our evidence that this theory of the greenhouse effect and its role in climate is correct. We'll end the first section of the course with "Global Recycling," a look at how the Earth recycles materials, particularly water and carbon, whose recycling is crucial to understanding how climate develops. That's the first half of the course.

The second half of the course is more about the human influence on climate. In *Earth's Changing Climate* part two, we'll start with Lecture Seven, "The Human Factor." We'll look at what we humans are doing in the context of the science that I described in the first half of the course, and we'll see how some of the balance that gives us the natural climate is being upset by changes we humans are making. We'll talk about computing the future. How do we project—not predict because it's not that accurate—but how do we project future climates? How do we understand what future climates are going to do? How do computer climate models work? and so on. We'll look in Lecture Nine at the "Impacts of Climate Change." What is climate change going to do to the planet besides the obvious, which is warming things up? What are some of the other impacts? Sea level rise, changes in the way species behave, changes in precipitation. What are some other impacts of climate change?

Then we'll move, in the last three lectures, to energy and climate issues because it's the human use of energy that, far more than anything else, is what's causing human beings to alter nature in a global sense for the first time; and in particular, to alter the climate. We'll look at energy and climate at length. We'll look at the resources we have to make energy for ourselves on Earth, and what some alternatives are to what we're doing today. In Lecture Twelve, we'll raise the question: Can we see our way to a sustainable future with a stable, benign climate? That's an outline of where we're going in this course.

There are a few other resources available to you that I strongly urge you to take advantage of. At the end of each lecture in your booklet, there is a list of suggested readings. I've chosen to break out the suggested readings into two different types. There are the suggested readings that you really ought to go into to remind yourself what this is about, to get corroborative evidence from other authors, and so on. Then I've given you a section of readings called "Going Deeper," where I give you books, and occasionally scientific papers, that go deeper into the issues I've discussed in that particular lecture. If you really are interested in looking quantitatively at the science behind those things, I urge you to look at those.

We get an awful lot of our information today from the World Wide Web, and so at the end of each lecture, I also give you websites to visit. Some of them are websites that give you more explanatory material on what I've covered in the lecture. Some of them are the websites that give you the raw data that I've used. If you wanted to make that graph that I showed you at the beginning of this lecture, of Earth's global temperature, I'll show you what website to go to that publishes that data, makes it available to the public, and you can go ahead and use that data yourself. A final reason for looking at the websites is that everything I say here is frozen in time, as of the moment I'm making this lecture. Those websites can update that material and keep you very much abreast of what's going on in this rapidly changing field. I hope you'll enjoy the rest of this course.

Lecture Two
Butterflies, Glaciers, and Hurricanes

Scope:

Rising thermometers aren't the only indicators of climate change. Other evidence points to increasing temperatures, as well as more subtle effects associated with changing climate. Hundreds of species show shifts in range consistent with a warming climate. Mountain glaciers and polar ice are melting, often at accelerating rates. Weather patterns are changing, with more intense precipitation events occurring, a narrowing of the day-night temperature difference, and increases in hurricane intensity. However, a variety of factors besides climate can influence biological, glaciological, and meteorological phenomena, and even different climate-related effects can work in opposite directions. This lecture also looks at these more subtle points and shows how statistical analysis nevertheless yields patterns that show clear "fingerprints" of climate change in a host of natural systems.

Outline

I. Surface temperatures aren't the only indications that Earth is warming.

　　A. Boreholes provide a record of surface temperatures in the recent past, because the surface temperature "signal" propagates into the ground. Borehole temperatures smooth out the year-to-year fluctuations and provide a general trend consistent with surface temperatures.

　　B. The heat content of the upper layers of the ocean shows an overall increase over the past few decades.

　　C. Temperatures in the lower atmosphere (the *troposphere*), although harder to measure than surface temperatures, show an increase consistent with the rising surface temperatures.

　　D. Temperatures in the upper atmosphere (the *stratosphere*) show a decrease. As Lectures Four and Five will show, this is consistent with increasing surface temperatures.

II. Earth's *cryosphere*—ice and snow cover—responds to changing temperature. Because it takes a lot of energy to melt snow and ice, these changes are relatively slow and reflect the overall temperature trends.

A. Most mountain glaciers around the world are shrinking at accelerating rates. Glacier growth and shrinkage are determined by a balance between accumulating snow and the effects of melting.

B. Arctic sea ice is melting at accelerating rates. Melting sea ice results in a *positive feedback*, as darker ocean water absorbs more sunlight and results in further warming. See Figure 2.

C. Land-based ice caps are also melting.

    1. Greenland's ice melt is especially significant and is accelerating rapidly.

    2. The situation in Antarctica is less clear, with melting at the margins and some ice cap growth in the interior.

    3. A concern is that meltwater flowing under the ice may lubricate ice sheets, making them prone to slide into the sea.

III. Tracking weather patterns over decades shows changes that would be expected with a changing climate.

A. There are more extremely hot days and heat waves.

B. Precipitation is increasing as warmer temperatures drive more evaporation.

    1. The change is not uniform, with some parts of the world drying and others becoming wetter.

    2. More precipitation is falling in intense, brief events.

C. The *diurnal temperature range* is decreasing, as night temperatures rise faster than daytime temperatures.

D. Hurricane intensity is increasing, especially in the North Atlantic. See Figure 3.

    1. Evidence for an increase in the number of hurricanes is elusive.

    2. Changes in hurricane intensity seem to correlate with rising sea-surface temperature.

IV. Plant and animal species show changes in geographical range and in timing of behaviors.

A. Northern hemisphere species are moving northward and upward.

B. Springtime events are occurring earlier.

C. Although these changes have complex causes, 87% are in the direction consistent with increasing temperature.

Suggested Reading:

Houghton, chapter 4, pp. 56–64.

Wolfson, chapter 14, section 14.2 to "Going Further Back."

Going Deeper:

Parmesan and Yohe.

Web Sites to Visit:

National Snow and Ice Data Center, http://nsidc.org/. Images and data showing changing conditions in Earth's cryosphere.

World Glacier Monitoring Service, http://www.geo.unizh.ch/wgms/. This Swiss site carries yearly data on hundreds of glaciers worldwide.

Questions to Consider:

1. Why might increased precipitation and cloudiness be expected consequences of a global temperature increase?

2. What are other factors besides climate change that might affect species ranges and account for the fact that most but not all species show changes consistent with increasing temperature?

Lecture Two—Transcript
Butterflies, Glaciers, and Hurricanes

Welcome to Lecture Two: "Butterflies, Glaciers, and Hurricanes." In the first lecture, we looked at the temperature trends over the last 150 years, and we found that the Earth has been, in fact, warming. We found that warming wasn't even, that things were roughly steady in the late 19th century, that there was a warming trend in the early 20th century, a roughly even trend in the middle 20th century, and then a much more rapid warming into the 21st century, from the last few decades of the 20th century. That's perhaps the clearest indication that Earth's climate is changing, but climate is about a lot of things other than simply temperature. It's about humidity; it's about moisture content of the soils; it's about cloudiness; it's about precipitation trends. It's about a lot of other aspects of the state of the atmosphere.

Furthermore, there are other indicators that can tell us about changing climate that are independent of the measured surface temperatures that we talked about in the first lecture. In Lecture Two here, we want to look at some of those other phenomena that may indicate the Earth's climate is also changing. Some of them, themselves, will be simply other ways of measuring temperature, but some will be entirely different phenomena—phenomena involving animal and plant species, phenomena involving storm intensities, phenomena involving ice—that also indicate that Earth's climate is changing. We'll be looking at all of those.

Scientists call some of these phenomena that are indicative of climate change "fingerprints"—they use the word fingerprints—of climate change. They use the word fingerprint in a particular way. Some of these changes that we are seeing—other than the thermometer temperature changes I discussed in the first lecture—some of these changes are themselves things that we would expect to have happen if climate change were occurring in a particular way for a particular reason. In particular, some of the changes we're observing today are changes we would expect as a result of anthropogenic—and that's the word I'm going to use again and again in this course, anthropogenic means of human origin; "anthropo," human. Some of the changes that we see are "fingerprints" of anthropogenic climate change, and give additional evidence to the statement I made in the first lecture—and the statement that I'll elaborate on in the second half of the course—that at least some of the climate change we've seen in recent decades is caused by human beings.

What are some of these fingerprints? What are some of these other indicators of climate change, other than simply taking the global temperature with thermometers? That global temperature record, remember, was surface temperatures. It was made from thermometers mounted at about the height of my eyes, above the Earth's surface, basically the surface layer of air. It was made from thermometers that measured marine air temperatures, temperatures just above the sea surface; and it was made from sea-surface temperatures, SSTs, temperatures of the surface waters of the ocean. It's temperatures at the surface layer of the Earth, but there are other temperatures you can measure that also say something about climate change.

For example, if you dig a hole down into the Earth, you'll find that temperature changes with depth. How should it change with depth? Well, if you think about it, the interior of the Earth is very hot; it's thousands of degrees. There's a molten outer core of the Earth, made largely of molten iron. Things are hot down there. That's the primeval heat that the Earth comes with. Some of it is still left over from the formation of the planet, and some of it comes from the decay of radioactive elements inside the Earth. I'll say more about that when we talk about energy sources because that is one of the energy sources available to us. If things at the surface were completely steady—if we had a steady climate at the surface—what we would expect to see if we drilled down into the Earth is an increase in temperature. It's not a very dramatic increase, but we would expect to see a measurable—and more importantly, steady—rise in temperature as we went down into the Earth.

What happens if the planet's surface is warming is that additional heat flows into the Earth from above, instead of just flowing to the surface from below. When that happens, the temperature profile that we would expect to measure if we drilled down into the Earth would give us warmer temperatures in the surface layers of the Earth than we expected. We'd have to drill down some significant distance before we'd begin to see that steady increase associated with the geothermal heat in the Earth's interior. That's a long-winded explanation of what's called *borehole temperature measurements*.

A borehole is simply a hole bored into the Earth, and we have lots of them. Every time we've bored for oil, we've made a borehole. There are abandoned oil wells, mine shafts, and all kinds of other boreholes around the world. If you study the temperature profiles in those, you're learning something about what's been happening to surface temperatures over time scales of decades to centuries, typically. Why is that again? It's because the surface temperature signal—if I can call it that—slowly propagates into the ground and leaves us a kind of record, with depth, of what the temperature has been doing. Until we get to that depth where the geothermal heat flow begins to dominate, the outer layers of the Earth's surface are dominated by these effects of recent climate change. We can read that climate change by measuring temperatures in boreholes. That's one way of getting at temperature change over the last decades to centuries.

Furthermore, those borehole temperatures are interesting because they smooth out fluctuations. As the heat gradually diffuses into the Earth, fluctuations—for instance, a particularly warm year or particularly cool year—those will be smoothed out, and we'll get a kind of average sense of what the climate has been doing over the years. What do the borehole temperature records tell us? They tell us simply that the Earth has indeed been warming. We don't have enough borehole temperature measurements to get a really clear global picture. But just about everywhere we've looked at borehole temperature profiles; the results are consistent with the surface temperature profiles I showed you for the 150-year temperature record. Boreholes provide a kind of smooth averaging that gives us a sense of what's been going on, on a time scale of decades or so. That's one example. A very recent study, by the way, suggests that the upper surfaces of the continents have absorbed an enormous amount of energy because of this heat flow from the warming climate. So, there's a big reservoir of energy, and that may, in turn, begin to have its own effect on climate. The land area, the land and the rocks, absorb this heat, and they can release that heat if they need to under certain conditions.

Another place where a lot of heat gets stored is in the upper layers of the ocean. Again, there are temperature profiles associated with the ocean in its natural state. As the surface begins to warm, some of that warming permeates down into the oceans, and it changes the temperature structure of the oceans. From that, we can calculate what's called the heat content of the upper layers of the ocean. That's a measure not of what the climate is doing right now, but what it's done over time, how much heat has accumulated in the surface layers of the ocean. That trend also shows an upward rise in temperature. The surface temperatures of the ocean have been rising—the temperatures in the top layer. The extent to which warm water goes downward into the ocean has been increasing, and there's an increase in the heat content of the upper layers of the ocean in the past few decades.

Finally, there's the atmosphere. The surface temperatures I talked about in Lecture One are measured, again, largely just a little bit above the surface—or in the case of sea-surface temperatures, in the upper layers of the water. They're not measured high in the atmosphere. What have those temperatures been doing? There's been a lot of discussion of that over the past few decades, and only in recent years has it become pretty clear what's happening there. In fact, if one looks in the lower atmosphere, the so-called troposphere, the lowest layer of the atmosphere that extends upward to a little bit higher than commercial jetliners fly, up to maybe eight or nine miles above the surface, 10 km; it varies quite substantially. That's the lowest layer of the atmosphere. That's the layer in which most weather occurs. Temperatures in the troposphere, in fact, show a warming.

Those temperatures are not easy to measure. The old way of measuring them was to send up weather balloons—so-called radiosons—equipped with suites of instruments that would then radio their results back. Of course, you were only probing a small part of the atmosphere, and only in a certain time. Those data are not terribly consistent, so it's difficult to build a global picture of the tropospheric temperatures. In recent years, we have instruments called microwave sounding units that are flown on satellites. They look down on the atmosphere, and they can calculate a rough average temperature for vast volumes of atmosphere based on microwave emission from oxygen molecules in the atmosphere. They're also a little bit hard to interpret because they can tend to smear out the temperatures in different layers of the atmosphere. There's been a lot of discussion of what those temperatures have been doing. In recent years, it's become clear that the troposphere has also shown a warming of approximately the same magnitude as the warming at the surface. Some of the latest measurements suggest the troposphere has, in fact, warmed a little bit more than the surface. If you looked at a graph of that, you would see a warming trend for the surface, and a slightly larger trend for the troposphere.

Above the troposphere is the next layer of the atmosphere, the stratosphere. In the stratosphere, things happen rather differently. The stratosphere shows a marked cooling. If you looked at a graph of the tropospheric and surface temperatures, and they showed an upward trend over the past few decades, the stratosphere would show a sharp downward trend; in fact, the more substantial downward trend. Wait a minute. How can I say the Earth is warming if the stratosphere has been cooling? As I'm going to show you in Lectures Four and Five, on the basis of the science of how climate is established, warming of Earth's surface, under the mechanisms we understand causing that warming—mainly the greenhouse effect, which I will elaborate again in later lectures—under the greenhouse effect one expects a warming of the surface to be accompanied by a cooling of the upper atmosphere. In fact, the temperature difference between the surface and the upper atmosphere is part of what drives greenhouse warming. If the Earth warms, you expect, ironically, that the stratosphere will cool. The more the Earth warms at the surface, the more the stratosphere should cool, and the more that temperature difference between the lower levels of the atmosphere and the surface, and the stratosphere. In fact, we see that the stratosphere has cooled substantially, and the troposphere has warmed, and so has the surface.

Earth has many spheres. It's got the biosphere, the sphere where life lives. It's got the atmosphere, the gaseous layer. It's got the lithosphere, the rocks. It's also got the cryosphere, as in cryogenic, cryostat; these words meaning cold. Earth has a cryosphere. The cryosphere refers to the ice and snow that exist at the surface of the Earth; some of it in the form of icecaps on places like Greenland and Antarctica, some of it in the form of floating sea ice, some of it in the form of mountain glaciers, some of it in the form of the seasonal ice and snow that falls in the temperate climates, and so on. The cryosphere.

The cryosphere: the ice and snow layers of the Earth, responds to change in temperature. As you know if you've ever left an ice cube out on the counter or something, it takes a while to melt ice. If you put an ice cube out on your counter in your 68°F house, and it doesn't instantly turn into liquid water. There's a lot of energy involved in changing the state of water from solid to liquid. It takes a long time; it takes a lot of energy. As a result of that, these changes occur slowly, and they therefore, like the borehole temperatures, are indicative of long-term trends in climate. They don't tell you what's happening from year to year or month to month. They look at long-term trends because it takes so much energy to melt ice, that the melting occurs over very long periods of time.

If you look at mountain glaciers around the world—take pictures of them from a few decades ago, take pictures of them in recent years—you will find that in many, many cases, you find substantial and dramatic retreat of glaciers. In the Cascade Mountains of Washington, for example, there are glaciers that, in the 1920s, filled entire valleys. If you take a picture of those glaciers today, you find they've retreated substantially up their valleys. There's far less area of ice; the ice is far less thick. Same thing in Alaska; for example, there are places where, in the 1940s, as recently as the middle of the 20th century, glaciers filled whole valleys. Today, there are lakes where those glaciers were, and way at the far end of the lake, you can see just the little bit of ice that's left of that glacier. Even tropical mountain glaciers—places like Mount Kilimanjaro that have substantial ice caps near their summits, even though they're in equatorial latitudes—those are diminishing rapidly, shrinking in area and thickness. The mountain glaciers around the world seem to be, by and large, shrinking.

Science is subtle, and there are other effects besides the warming of the planet that affect some of these things. What a glacier does is determined by a balance between its melting or its "calving off" of ice at the point where it reaches the sea, if it does, and the accumulation of snowfall on the glacier. One of the effects of climate change may be to cause more precipitation, and so some glaciers, depending on exactly how they're situated, may actually grow under global warming because they get more precipitation falling as snow and adding to the glacier. These things are subtle. A glacier that is bucking the trend of most mountain glaciers to be shrinking is not a counter-indication. It says let's look at this glacier and understand what its particular circumstances are.

If you do look at a whole lot of glaciers—if you were to make a graph, for example, of what a whole lot of glaciers have been doing over the past decades, or even a century or so—you would find that most mountain glaciers' lengths remained relatively constant until a few decades ago, and then most—again, not all—began declining in length of the glacier, which is one measure of a glacier's extent. Another measure is its area; another measure is the thickness, another measure is the volume of the ice. But if you simply looked at the length of glaciers—and scientists have done this, tracking glaciers all around the world—you will find glacier lengths have tended to decline. The decline has been particularly steep in recent decades, time that coincides with that steep increase in world temperature that I showed you in Lecture One. Most mountain glaciers are retreating.

A very different kind of ice is arctic sea ice. Arctic sea ice is floating ice in the Arctic Ocean. Of course, in the Arctic, we have ocean at the North Pole; we have a continent at the South Pole. So the two arctic regions, the Antarctic and the Arctic, are very different geographically and climatologically. In the Arctic, the floating sea ice is shrinking substantially. If you, for instance, look at pictures of Arctic sea ice, and you look particularly in September, which is about the time the ice has reached its minimum area, you will see again, over a time scale of only 20–30 years, substantial decreases in the extent of the Arctic sea ice over that time, the minimum ice cover.

Take a picture based on satellite imaging, and the minimum ice cover is substantially lower at the present than it was 20–30 years ago. The Arctic sea ice is diminishing in extent. It's also diminishing in thickness. We know this from declassified information from nuclear submarines that ply their way underneath the Arctic ice, and did so in the 1950s and '60s. They got measurements of the thickness of the sea ice from below, and we know that the thickness of the ice is also decreasing. If you were to make a movie, put together a bunch of data and make a movie of it, you would see the Arctic sea ice gradually decreasing. You would see that some years it grows, some years it shrinks; but overall, the trend is a decrease in the Arctic sea ice in the extent of the Arctic sea ice. That trend shows you, as the ice blows around and expands sometimes and detracts sometimes, that overall, we're losing ice. Overall, something is happening to supply enough energy to turn that ice into liquid water.

I am going to show you one graph here—and for you audio people, this graph is in your booklet, so take a look at it. These are measurements of the aerial extent, the area, of the Arctic sea ice over time since the 1970s. Measurements are taken every year at what's called the minimum ice extent. It typically occurs in late summer or early fall, typically sometime in September. This graph shows the measures of that area, and the area has dropped substantially, something on the order of 30% or so, in the time since these accurate measurements were taken in the 1970s. It's a steep downward trend. It has accelerated somewhat in recent years. Again, as with all scientific data, there's some fluctuations involved there, but the area of Arctic sea ice has been decreasing substantially.

You may have heard a lot about ice in Greenland. Ice in Greenland represents a different kind of ice. That's an Arctic ice cap; that's ice that is on land, land-based ice. In Greenland, ice has been melting at accelerating rates quite recently. There's been a bit of alarm about this. You get the sense that perhaps a large amount of Greenland is melting. What we're measuring typically is how much of the surface layers are beginning to show melting, rather than staying solid all year round. Those numbers are increasing dramatically. We aren't losing an enormous amount of ice yet, but the trend is upward, and it looks like Greenland is beginning to lose ice at a significant rate. The Greenland ice cap is melting.

The situation in Antarctica is a lot less clear. At the fringes, Antarctica is showing the effects you would expect from global warming, but the interior of Antarctica, being a very cold, dry, desert-like environment, if any precipitation falls in the interior, it doesn't melt. Even with substantial warming of the planet, the interior of Antarctica is nowhere near the point where it melts. It actually is accumulating snow. If there's more precipitation associated with global climate change, Antarctica may actually increase in the interior extent of its ice sheets or ice caps, but the regions near the margin of that continent are definitely showing the effects of warming.

One particular concern we have with places like Antarctica and Greenland is, as the ice begins to melt, melt water can flow not only on the surface, but it can flow down through crevasses, and it can flow along on the bottom of the ice sheet, between that and the rocks, and it can lubricate that region. Where before, the ice sheets may have been locked rigidly to the rocks, now that gets lubricated, and whole large pieces of ice sheet could then slide off the continents and into the ocean. There's worry that that could be something that could precipitate an abrupt rise in sea level. It doesn't appear to be a substantial possibility in this century, although it could happen, and that's one of the worries associated with this melting.

There are other indicators, other fingerprints, of global climate change. For example, if you track weather patterns over a few decades, there are changes. Again, as I said in the first lecture, weather is different from climate. Weather is the day-to-day stuff that's happening in the atmosphere. But if you look at the trends in weather over the decades that tells you something about what's happening to climate. What are some of these changes? Well, you would expect there to be more heat waves and prolonged hot spells if it's warming; that's sort of obvious. Although what's perhaps not obvious is why those heat waves and prolonged hot spells should rise dramatically if the temperature, on average, changes only a degree Celsius or something. I'll show you in a later lecture exactly why that's the case. It's the same reason, by the way, that I showed some of you in my video course *Physics in Your Life* about why food spoils in the refrigerator if the temperature goes up only a little bit. It's the same phenomenon. A very small change in the average temperature can dramatically shift conditions at the extremes, so we expect to see more extremely hot days and more heat waves.

Precipitation is changing; the patterns are changing. In particular, the precipitation globally seems to be increasing. There's a slight increase in precipitation, but some areas are drier. That change is not even. Some areas are getting drier, and some areas are getting wetter. There's more precipitation coming in brief intense events. That's another characteristic fingerprint of anthropogenic climate change, and we're definitely seeing more precipitation coming in brief intense events. If you think about that, you might say well, if you were a plant, or if you were some soil that was going to erode, how the precipitation comes—a gentle rain that lasts an hour, a brief downpour that lasts 10 minutes—that makes a big difference, even if the amount of rainfall stays unchanged. Here's a more subtle change in what's going on, and that's a change that, again, is a fingerprint of anthropogenic climate change, and there's pretty substantial evidence that we are seeing more precipitation coming in brief intense events than we used to. Again, there's evidence that we are seeing more precipitation generally globally; not a whole lot more, but more, and we are also seeing some parts of the world drying out, and some parts getting wetter.

Another thing that's changing is the diurnal temperature range, what climatologists call the DTR, the temperature difference between night and day. That's decreasing. It's actually warming more at night than it is in the daytime, and that's causing the temperature range from night to day to decrease. That's probably associated with increased cloudiness. In my home state of Vermont, there's a big worry about that because maple sap production requires very cold nights and then warm spells in the day. If we lower the diurnal temperature range, we don't get those conditions, and in fact, the maple syrup industry may leave my state for Canada as a result of global climate change.

We've heard a lot about hurricanes in recent years, and we've heard a lot about whether hurricanes may or may not be related to global climate change. There, the issues are quite subtle. There are definitely some changes in hurricanes. I'm going to show you another graph, this is a graph that is associated with the North Atlantic and with the intensity of hurricanes as measured by an average of the actual power contained in the winds of the hurricane multiplied by the time over which those winds are blowing, which gives you a sense of the sort of total energy contained in the hurricanes. That's done over a whole hurricane season. This graph shows you two curves. The lower curve is the sea-surface temperature in the North Atlantic; the upper curve is the hurricane intensity as measured, as I just described, by that total energy associated with the power contained in the winds multiplied by the time over which those winds were blowing.

For those of you who are mathematician types, by the way, that power in the wind goes as the cube of the wind velocity. That's because kinetic energy goes as the square of the wind velocity, and therefore, the wind is carrying that kinetic energy along at its velocity. The whole thing goes as the cube of the wind velocity. If you double the wind speed in a hurricane, the intensity of that hurricane goes up eightfold. Here you see, in the North Atlantic at least, quite a clear correlation between the intensity of the hurricanes and the sea-surface temperature. There's plenty of good reasons for expecting that warm sea-surface is what drives intense hurricanes. That's not surprising, and that seems to be something that's related to climate change.

In the interest of scientific accuracy, I will tell you if you do the same thing—and it has been done for other ocean basins—the correlation is less dramatically obvious. One thing that is not at all obvious is whether to expect an increase in the number of hurricanes, or whether, in fact, we've seen an increase in the number of hurricanes. It's pretty clear we've seen an increase in their intensity, but have we seen an increase in their numbers? These things are a little bit harder to get at than you might expect. Something like hurricanes presents a kind of ambiguity for us. We can get at some aspects of hurricanes that may be related to climate change, but hurricanes are subtle, and we don't know all the details.

Finally, let me end with something related more to life itself, and that is what happens to species—and that was the butterflies in the title of this lecture—what happens to species as a result of climate change. A lot of things happen to species. Species undergo a lot of behaviors that, in fact, are related to timing of when the spring arrives, what the highest temperatures are, when the first frost is. Things like that affect plant and animal species both. The results I'm going to talk about here are from not one study of a particular species of butterfly, or bird, or tree, but rather of a "metastudy"—as they're called—that looked at hundreds and hundreds of studies of individual species and came to kind of general average conclusions about what was happening.

What do those species do? In the Northern Hemisphere, on the average, species seem to be moving northward. Their ranges—the southern extent of their range and the northern extent of their range—seem to be moving northward at about six kilometers per decade. Every 10 years, if you take a species whose southern range was New York City or something, then 10 years later, its southern range is six kilometers, about four miles, north of there. The same goes for the northernmost range. Again, this is not about any one species. It's kind of an average over many, many species, but there you are when you do that average. Mountain species, species that live in high altitudes, tend to be moving upward. That's the only way they can move easily. They're moving upward at about six meters per decade, so, six kilometers per decade roughly for the species that are moving toward the poles, and roughly six meters per decade for the species that live in the mountains.

Other events, spring events, seem to be taking place earlier by roughly two to three days for every decade. Every 10 years, a lot of events that you think of in spring move earlier. I know, having lived in Vermont quite a while, I used to plant my tomatoes on Memorial Day; now I plant them in early May. That's an example of that kind of spring event moving earlier by two to three days per decade. You can see it's not always a bad thing. Some of those events include nesting, when birds build their nests, when animals breed, bud bursts, the bursting of buds to make leaves, when things flower, or when the spring migration ends. All those things are happening earlier by about 2.3 days per decade. There are some other examples of things that species are doing that are varying.

I want to emphasize that something as complicated as the behavior of a plant or animal species is not determined solely by climate. There are other factors. What are other species doing? What are the prey or the predator species doing? What are diseases that affect that species doing? There are all kinds of other effects. These are subtle scientific questions, and that's why this metastudy of hundreds and hundreds of different species gives us a sense of what's happening overall. We don't expect every single species to be doing that. In fact, if you look at these metastudies, you will find that roughly 87% of them have changes that are consistent with what you'd expect if those changes are driven by global warming.

A small fraction of them, species whose behaviors may be dominated by things other than the actual warming associated with climate change; maybe they're affected by precipitation. Maybe they're affected by other species. This is not surprising; about 13% of species in this metastudy of hundreds of species have shown changes that are in the opposite direction than you'd expect associated with a temperature change. That's simply an indication that what controls the behavior of a species is a lot more than climate alone, and that's not surprising. But the main point to take away from these studies is that, of the species studied, 87% showed changes in their behavior, changes in their range, their mating behavior, their early spring behavior, their migrations, etc., that seem to be consistent with global warming, and only 13% showed changes that bucked that trend.

Let me end by summarizing what we've seen here. We've seen a whole lot of fingerprints for global climate change in a whole lot of different areas. The temperatures in boreholes, measuring the temperature down into the Earth, and indicating that historically there's been a flow of heat into the Earth, and therefore the temperatures are rising at the surface. There's a change in the temperature occurring. We've seen the heat content of the ocean rising. We've seen the atmospheric temperature rising in the lower atmosphere, but falling in the stratosphere, which is something, as you'll see in a few lectures, that is exactly what we would expect if there was greenhouse effect global warming. We've seen changes in the cryosphere, which are largely associated with melting and shrinking of polar ice, mountain glaciers, ice caps, and so on. Finally, we've seen changes in the behavior of species of with respect to the ranges they occupy, and with respect to the timing of characteristic behaviors like mating, and nesting, and bud bursts, and so on. There are many fingerprints that point to the fact that Earth is indeed warming, that the climate is indeed changing.

Lecture Three
Ice Ages and Beyond

Scope:

Thermometer-based temperature records go back only 150 years. This lecture explores the techniques scientists use to push global temperatures back thousands, millions, and even billions of years. In particular, we'll explore the cyclic pattern of ice ages punctuated by brief warmer spells. Our understanding of this pattern involves a complex interaction among astronomical effects on Earth's orbit and axis, along with feedback effects that drive changes in the levels of atmospheric carbon dioxide and other gases. The picture of past climates developed in this lecture will be important in the next few lectures, where we seek to understand the scientific principles that establish climate, and in later lectures, where we show why recent climate change is largely attributable to human activities.

Outline

I. Are the patterns of variation seen in the 150-year instrumental temperature record typical of natural variations in Earth's climate? Or is there something unusual about recent temperature increases, especially in the past few decades? Answering these questions requires pushing the temperature record back in time.

   A. Because there aren't enough accurate temperature-based measurements to calculate a global average temperature before the mid-19th century, scientists use *proxies*, physical quantities that serve as indicators of temperature. Commonly used temperature proxies include:

   1. Annual tree rings, whose thickness and wood density varies with temperature and length of growing season. Tree rings are most valuable in regions showing large winter-summer temperature variations.

   2. Coral reefs, which form annual layers of calcium carbonate whose analysis yields information about the temperature at the time of formation. Corals are most useful for tropical ocean temperatures.

   3. Lake sediments, the thickness of which indicates the rate of snowmelt that feeds streams carrying sediment and, thus, provides a measure of springtime temperatures.

Further, the pollen content of lake sediments tells us about plant species, giving a general indication of climatic conditions.

4. Isotope ratios from ice cores and shell-forming marine organisms also provide a measure of temperature, specifically, ratios of the two stable isotopes oxygen-16 and oxygen-18; a similar technique uses hydrogen isotopes.

B. A number of independent multiproxy studies use statistical methods to combine different proxy temperatures to reconstruct global or hemispherical temperature patterns back one or two millennia. See Figure 4.

1. The temperature reconstructions vary somewhat, but all show patterns of gradually declining temperature over most of the past millennium, followed by a sharp upturn through the 20th century and into the 21st century.

2. These reconstructions suggest that Earth's current climate is the warmest in at least the past millennium.

3. An independent study by the U.S. National Academy of Sciences affirms these findings, with considerable confidence in the reconstruction of the past 400 years and more uncertainty for earlier years.

C. Ice cores from Greenland and Antarctica use oxygen and hydrogen isotope ratios to provide temperatures ranging back nearly 1 million years.

1. These records show a cyclic pattern of brief (10,000- to 20,000-year) warm spells called *interglacials*, separated by longer cold spells (*ice ages*). For the past half million years, this cycle has repeated on roughly 100,000-year intervals. See Figures 5 and 6.

2. This pattern results from subtle changes in Earth's orbit and tilt, along with complex feedback effects in the climate system.

II. Climate reconstructions going back still further use a variety of proxies, including geological evidence for ice, fossil vegetation, and oxygen isotope ratios from fossil plankton.

A. This evidence suggests that Earth has been considerably warmer in times past, especially during the age of the dinosaurs, hundreds of millions to tens of millions of years ago.

B. There is also evidence of as many as four *snowball Earth* phases, when the planet may have frozen solid. These occurred between 750 and 580 million years ago and were followed by rapid thawing associated with changes in the atmosphere due to gaseous emissions from volcanoes.

Suggested Reading:

Houghton, chapter 4, pp. 64 to end of chapter, emphasizing temperature.

Wolfson, chapter 14, "Going Further Back" to end of chapter.

Going Deeper:

Ruddiman, chapters 7–10.

Harvey, chapter 1.

IPCC 4, chapter 6.

Web Sites to Visit:

Carbon Dioxide Information Analysis Center (CDIAC), http://cdiac.ornl.gov/trends/ temp/contents.htm. At this particular CDIAC page, you'll find graphs and data on global and regional temperatures from sites around the world, going back hundreds of thousands of years.

Questions to Consider:

1. Distinguish between an instrumental temperature record and a proxy reconstruction.

2. What is the approximate temperature difference between interglacial warm periods, such as we enjoy today, and the ice ages? Express in both Celsius and Fahrenheit.

3. What is the ultimate cause of the cyclic pattern, seen for the past half-million years, whereby ice ages alternate with briefer warm spells?

Lecture Three—Transcript
Ice Ages and Beyond

Lecture Three: "Ice Ages and Beyond." In the first lecture, we looked at the 150-year temperature record from thermometers and saw that Earth has indeed been warming, especially in recent decades. In the second lecture, we looked at a number of so-called fingerprints, other phenomena that also are indicative that Earth's climate is changing, either because temperature is rising or because other climate-related phenomena—like species ranging, and so on, melting of ice—are occurring. But we'd really like to know whether that roughly 150-year record of temperature, corroborated by these other phenomena, is just a natural occurrence in the history of Earth's climate, or whether there's something abnormal, something unusual, particularly about the warming of recent decades.

What we need to do to answer that question is somehow to go back further than that 150-year time horizon where we have enough data from thermometers to tell us something about Earth's global temperature. The problem is, of course, that we don't have thermometers that were in enough quantity before about 1850, and spread enough around the globe, to measure an average global temperature. A few hundred years before that, we don't have any kind of temperature measuring instruments at all, so how do we get back further? We've got to do that if we're going to find out whether the climate change we're seeing now is, in fact, something unusual or something just to be expected in the history of the planet.

What we have to do is find measurements that serve as so-called proxies for temperature; other things that we can measure that kind of stand in for temperature. Sometimes they may also stand in for other aspects of climate or other phenomena all together, but at least they have to be able to give us some indication of what temperatures were doing at times back before that historical temperature record. It would be really nice if these proxy temperature measures also extended into the region where we have thermometer temperature measurements because then we could cross-correlate them and understand them. What are some of these proxies we use to go further back in time and try to establish so-called paleoclimate, climate in ancient times, or even climate in fairly recent centuries? What do we use?

There are a number of common proxies; one of them is tree rings. The oldest trees go back thousands of years. The rings in trees, the annual growth rings, tell us a lot of things. They tell us about conditions when the plant was growing. They tell us about temperature. They tell us about the length of the growing season. They tell us about humidity. They tell us about where the tree is in its age cycle. All these things have to be sorted out. But if you look at tree rings, you look at the interior structure of a tree that's been sawed in half, you see these annual growth rings laid down as the tree undergoes its annual spurts of growth. The rings vary in thickness, they vary density of wood, they vary in other quantities that tell us something about climatic conditions during the years that tree was growing. Because tree rings are laid down annually, we have a precise measure of when a given tree ring is associated with, and therefore we have a real dateable record of past climate, as measured by tree rings.

Tree rings are most useful in regions like the temperate zones, where there are large summer/winter temperature differences. Tree rings in the tropics aren't going to tell you a lot, but tree rings in the temperate climate, where there's this large differential between summer and winter, tell you quite a bit about the climatic conditions at that time. If you go into the tropical oceans, you find, in contrast, coral reefs. Coral reefs form annual layers of the calcium carbonate, which is what forms the shell structure of shelled marine organisms. An analysis of those layers tells you something about the temperature of the water at the time the coral reefs were forming. Coral reefs are analogous, in some sense, to tree rings, except they work for warm tropical waters, whereas tree rings tend to work for temperate land areas. They're complementary in that sense. We can get at two different regions of the planet by looking at tree rings and looking at coral reefs.

Lake sediments. Freshwater lakes have layers of sediment, typically, on the bottom. Where do those sediments come from? They come from streams that are washing the sediment in. There tends to be a lot of stream flow in the spring when there's snowmelt going on. Then as the summer progresses, the stream flow dies down; sometimes it goes away all together. As a result of that, the sediments in freshwater lakes tend to be deposited in layers, which you can again—as with tree rings or coral reefs—identify on a year-to-year basis. You actually look at the sediments. You take a core of them, and you can see individual layers. The thickness of those layers tells you something about the intensity of the snowmelt. They tell you how rapidly the streams were flowing. They tell you a measure of the springtime temperatures, roughly. Also in those sediments, one often finds pollens. If one goes back long distances in time, then those pollens tell you something about what kinds of species happened to be living in the surrounding land. Lake sediments provide another proxy indicator of climate.

Perhaps the most significant one, and the one that's going to take a little bit more scientific explanation here, are so-called isotope ratios. I need to get a little tiny bit into nuclear physics here. I talked about some of this in my course *Physics in Your Life*, and I'm going to talk about it again a little bit here. What is an isotope? An isotope is one particular version of a chemical element. What's a chemical element? A chemical element is a substance characterized by a particular atom. In particular, it's the number of protons—the positively charged subatomic particles that make up atomic nuclei, along with neutrons—that determines what kind of a chemical element something is.

The reason the protons determine that is that the protons are electrically charged, positive. They attract an equal number of electrons to complete the atom; the electrons orbiting around the nucleus in the simple classical picture of an atom. It's that number of electrons, particularly the outermost electrons and their configuration, that determines the chemical behavior of the atom. The individual chemical elements, and their very different chemical behaviors—for example, oxygen versus hydrogen, versus gold, versus uranium—those different chemical properties are determined by the number of protons in the atomic nucleus. The number of neutrons doesn't affect the chemical properties because neutrons are electrically neutral. So, you can have different so-called isotopes of the same element that have different numbers of neutrons, but the same number of protons. Chemically, they're essentially indistinguishable. Two isotopes of oxygen are both oxygen. You can breathe chemical oxygen with either isotope, and it will work pretty much the same in your system, for example.

But the two isotopes differ, and they differ in mass because an isotope with more neutrons is heavier, so their physical properties are slightly—typically not dramatically—different. That's what isotopes are. I want to look, in the context of this course and temperature proxies, particularly as isotopes of hydrogen and oxygen. Hydrogen a very common element, a component of water that normally consists of just a single proton. It's the simplest element. It's one proton in its nucleus, surrounded by one electron; that's what hydrogen is. But there are two other isotopes of hydrogen. One is called deuterium; its nucleus consists of a single proton—that's what makes it hydrogen, one positive electric charge—but also one neutron. Because a neutron and a proton weight about the same, in fact, deuterium is about twice as heavy as ordinary hydrogen. Deuterium is a stable isotope, which means it can exist basically forever, and there is some deuterium among the hydrogen here on Earth and elsewhere in the universe. In fact, about one out of roughly every 6,500 hydrogen atoms is, in fact, deuterium. It's also called H_2 because it's got two nucleons—two either neutrons or protons—in its nucleus. So there's deuterium, and it is twice as heavy as hydrogen, although there isn't very much of it.

Oxygen, another common material on Earth—in fact, the most abundant material in Earth's crust is oxygen—comes commonly in the isotope oxygen-16. Oxygen has eight protons in its nucleus. Any nucleus with eight protons is oxygen. It's got eight electrons surrounding it to make an atom, and that determines its chemistry. Most oxygen is oxygen-16, but a small amount of oxygen is oxygen-18. Oxygen-18 has two extra neutrons; there's also oxygen-17. There's also, by the way, a H_3, but that's radioactive, so it doesn't tend to hang around. But the oxygen isotopes 16, 17, and 18 are stable. They can exist basically forever, and 16 and 18 are particularly significant in trying to measure temperatures, as are the hydrogen isotopes, hydrogen and deuterium. So what's the deal? How do we figure out what temperatures have been based on these isotope ratios?

There are a number of ways we do it, depending on what we're looking at. I want to focus on looking at ice cores. In the Arctic and Antarctic, we can drill into ice sheets that are a mile or so thick. Again, the ice tends to form in layers associated with the annual snowfall. Sometimes we can see individual years in the ice core record, and other times those are kind of smeared together, the dating is a little bit more approximate. But in any event, we can go back in these ice cores, and we can look at the ratio of the two isotopes of, say, oxygen in the precipitation that fell and eventually formed that ice. What do we do? Most oxygen, again, is oxygen-16, the lighter isotope—eight protons, eight neutrons. But a little bit of it, about 0.2%, is the heavier oxygen-18. Again, eight protons because it's still oxygen, but in this case 10 neutrons, to give it a total of 18 nucleons in its nucleus.

These things are basically similar chemically, but again they're different physically, and the oxygen-18 is heavier. That means when water containing oxygen-16 is near the surface and is warmed, the oxygen-16 containing water evaporates preferentially because it's lighter. It's more likely to go flying out into the air than the heavier isotope. What that already means is that oxygen-18 is depleted in the water that goes into the atmosphere. Atmospheric water vapor is somewhat depleted in oxygen-18. Why? Because the O_{18} preferentially remained behind when the water evaporated because its nuclei are heavier. The water that's in the vapor in the atmosphere is depleted; it has less oxygen-18 than the natural abundance of oxygen-18.

What else happens? Water typically evaporates in the tropical regions where it's warmest, and it's transported into the temperate zone, and into the Arctic zone, by the motions of air that is one of the major conveyors of energy from the tropics to the poles. If we didn't have that conveyance, by the way, the poles would be a lot cooler relative to the equator than they are. But there's a transport of energy associated with the transport of evaporated water from the equatorial regions. However, in the process of going from the equator toward the poles, the oxygen-18, containing water, being heavier, tends to precipitate out more readily. First, you evaporate water, and it tends to be depleted in oxygen-18. Then, as the water moves pole-ward, it gets further depleted. How much depleted it gets, you can see, depends somewhat on the temperature. The cooler it is, for example, the more the oxygen-18 will precipitate out, and that leaves Arctic and Antarctic precipitation even further depleted in oxygen-18. The cooler the climate, the sooner the oxygen-18 is gone, and so the more depleted is the precipitation that eventually falls in the Arctic.

That precipitation falls typically as snow. It eventually is compressed into the ice that makes up the ice sheets, and it preserves, as I said, either an individual yearly layering, or in some cases, just a smeared-out layering that allows us approximately to date it. But if we go in and core that ice, pull out an ice core, we then have a dateable—

sometimes dateable by year, sometimes dateable by rough era—record of the temperature at the time that precipitation fell. A similar technique works the same way with the hydrogen isotopes because, again, there's water that contains H_1, and then about one in every 6,500 waters contains an H_2, a deuterium atom. Either oxygen isotopes or hydrogen isotopes can be used to date particularly ice cores. They tell us something about conditions when that ice fell. Unlike the global average temperature that I talked about in the first lecture, they tell us about conditions at particular places; in Antarctica, in the Greenland ice sheet, or whatever. We can infer from those something more general about the temperatures on the Earth.

What do we do with these different proxies; the tree rings, the coral reefs, the isotope ratios, and so on? We put some of them together and make multi-proxy studies, studies that take a lot of these proxies into account, and do a detailed statistical analysis and try to reconstruct temperatures back into the past. The rest of this lecture, I'm going to spend primarily looking at some of those reconstructions of temperature. I'm not going to call them measurements because we haven't actually measured temperature. We've attempted to reconstruct temperatures based on these proxies and our well-founded scientific understandings of how these proxies reflect climatic conditions, including temperature, at the times that the tree rings were formed, or the precipitation fell, or whatever.

The first set of proxy studies I want to talk about are studies that push us back roughly 1,000 years. What these proxy studies suggest is that Earth's climate—to get to the bottom line quickly—is that Earth's climate at present, and in the past few decades, has shown a warming that is unprecedented on the 1,000-year time scale. I want to show you a graph here—actually, I want to show you a lot of graphs here; there are actually 11 graphs I want to show you. These graphs represent 10 different reconstructions of the climate of roughly the past 1,000 years. Every wiggly line you see is a different scientist or group of scientists' different study of what the temperature probably looked like on Earth back over the last roughly 1,000 years.

I said there were 11 lines, there's only 10 of those wiggly lines. There's a thick line that occupies only the rightmost portion of the graph and extends highest right at the rightmost edge, and that's the same instrumental temperature record that I showed you and talked about in Lecture One. Now it's been crammed into this very short piece of the overall 1,000-year graph, and so you see that rise in the early 20th century, that fall in the mid-20th century—a slight fall that's exaggerated here—and then you see that very steep rise at the end of the 20th century. You see all these wiggly proxy studies. They are not measurements of temperature, but reconstruction of temperature, and they all show roughly the same thing.

There's some variation, and you can argue about that and talk about that, who's right and who's not right. But they all show basically a slight downward trend over roughly the first 900 years of the millennium, and then a fairly steep rise at the end. If you had to summarize that in one quick sentence, that's how you would describe the graphs. If you were a climatologist, you might want to go into more detail and talk in more detail about what's going on there. But the basic big picture is, over the last 1,000 years, there's been a very gradual decline in temperature, followed by a sharp upturn at the end of that period in roughly the 20th century, late 19th through 20th century. On that basis, we can say that the temperature rise we've seen in the last few decades has been unprecedented on time scales of at least 1,000 years. How good are these reconstructions? That's a little hard to know. One has to assess them; one has to look at the statistical methodology used.

But again, we're looking at 10 different studies—10 different independent studies—that all point to basically the same thing. The U.S. National Academy of Sciences did a study of these 10 studies and concluded that, at least for the last 400 years, we can have considerable confidence in these studies. Going further back, the uncertainties get bigger, but the general trends are still basically the right thing. These are not the final word on what temperatures have been doing on this planet over the past 1,000 years, but they almost certainly give us a pretty good picture of the overall big picture. The overall big picture conclusion is that the temperature rise of recent decades, since about the 1960s or 1970, has been unprecedented, at least on a 1,000-year time scale. That's the beginning of the answer to the question I started this lecture with; is the temperature rise we looked at in Lecture One, augmented by these fingerprints, unusual in the history of the Earth's climate? Over 1,000 years, it looks like it probably is unusual.

What happens if we go back further? We can go back a lot further using ice cores from either Greenland or Antarctica, and there are a number of studies viewing these. I'm going to show you some results from drilling at the Vostok Station in Antarctica; it's a Russian site. That climate record has been available for quite a few years. There's a newer study that goes back almost a million years, the Vostok study goes back about half a million years. There's a newer one now that goes back almost a million years, by a European group. I'm going to show you the results from the Russian Vostok area study. This graph is showing a picture of the climate of the past 420,000 years. You're seeing on the horizontal axis thousands of years before present, with the present at the right, zero years before the present, going back to over four hundred and something thousand years. You're seeing temperature deviations on the vertical axis—these are measured in degrees Celsius—from the present-day temperature.

You'll see immediately that there is a pattern to this picture. There's a pattern of relatively short, 10,000 to 20,000 years in duration, warmish spells when the temperature is at or about the present-day temperature. Again, the present-day temperature on this graph is zero. It's just an arbitrarily chosen to measure deviation from the present-day temperature. If you look at zero, zero represents the present-day temperature. Some of those spikes go a little bit above the present-day temperature by a few degrees, but most of the time is spent in a much lower temperature state. You'll also notice that we rise rather quickly into those warm spells, and then we fall rather slowly into the cool spells, and we rise again. The time period between these warm spells is about 100,000 years, roughly. So for the past roughly half million years, this pattern doesn't go back, unchanged like this, too much longer. But for roughly the past half million years, the Earth has been in a cyclic climate pattern in which the temperature has mostly been at a fairly low temperature state, and then it's risen, for 10,000 years or so, to a warmer state, stayed in that state for a fairly short time, and then dropped down again, and again with roughly 100,000-year periodicity.

Now let's take a look at a close-up version of that. These cool periods are ice ages; the warm periods are called interglacials. I want to focus now on the most recent cycle from the last warm period before the one we're in now, and the ice age in between. Here's the most recent cycle. It's the same data, same temperature deviation from the present in degrees Celsius. The cool period is the ice age, about 120,000 years ago—I said the period to see was about 100,000 years—about 120,000 years ago, the period to see there was a warm spell. We're now in another warm spell. We're not quite at the warmest time in that warm spell, at least not yet, but there we are. A number that I would like you to walk away from this course from is roughly the difference in temperature between that warm spell that we're in now—or one of these typical warm spells—and the temperature sort of typical of the ice age.

What is that difference? You can argue; is it 10°, is it 4°? It's not 1°; it's a lot bigger than 1°. It's not 100°; it's not 20°. I would call it 6°; you could call it 8°, sort of getting it typical here. But roughly 6°C separates the present climate from that of the ice ages when North America, all of Canada, basically, and the northern United States had a mile or two of ice setting on top of them. When I said in the first lecture that a few degrees can make a big difference, I meant it. A few degrees in the global average temperature can be the difference—6°, in the global average temperature—can be the difference between the present conditions and an ice age when there was two miles of ice atop Montreal, for example, or northern Vermont, or whatever. That's a significant difference. When you hear people talking about climate change of a few degrees, the global average climate change of a few degrees is climatologically extremely significant.

What causes this change in Earth's temperature, this cyclic change? We don't understand all the details yet, but what we do understand is that this temperature change is not caused by any one particular factor. It's probably triggered by astronomical factors related to Earth's orbit. Those astronomical factors are probably things like the tilt of the Earth's orbit, the variation in the ellipticity of the orbit, how elliptical is the orbit, the changes in the tilt of the axis as the Earth goes around in its orbit change, how much of the time is spent in winter versus summer. All these things may trigger very subtle changes. What we believe then happens is that those subtle changes cause feedback effects in the Earth's climate system, feedback effects that I'll talk about a lot more when we deal with climate models. We will see just exactly what we think causes these effects. I'll say a little bit more about that in subsequent lectures, and look in more detail at those orbital effects that we think cause this to be triggered.

Again, these orbital effects are subtle; they are very small effects. Then they trigger changes within the climate system that rapidly build up the temperature into one of these warm spells. Then gradually, as the orbital parameters change slightly, we gradually drift back down into that low temperature state we call the ice age. Don't think of this as a clear indication that there is some dramatic effect causing the temperature change. No, there's some subtle effect that causes the beginnings of a temperature change. That temperature change then leapfrogs through various feedback effects in the Earth's climate, and away we go.

Can we go back further? Yes, we can. It's harder to go back further with certainty. We're quite certain about the temperatures I'm showing with these hydrogen isotope graphs I've just showed you, based on the hydrogen isotope temperature data. But can we go back further? Well, we can, and we have other proxies for that, and some of the same proxies we've been talking about. But we can look, for example, at oxygen isotope ratios in marine organism shells that are trapped in ocean sediments. Again, sediments form in layered structures, so we have a sense of what the time is that they were formed at. We can measure the oxygen ratios that tells you something about climatic conditions at the times those shells were formed, which then sank to the bottom of the ocean and became part of the sedimentary—ultimately of the sedimentary layer, or the sediment layer at bottom of the ocean. That's one example.

Another thing we can do is look at fossils, look at fossil leaves. Again, fossils occur in rocks. The rocks are layered; the layering is a rough indication of age. So, you can look at fossil leaves, and just the type of leaves you see tells you something about the climate. Was it tropical? Were these deciduous leaves? Was this a more temperate climate? You can begin to learn something about climate from that. You can also learn something about the volume of ice. There are a number of indicators that give you the volume of the ice. You can tell where the sea level was high or low. If sea level was low, that indicates that a lot of water was locked up in ice. If a lot of water was locked up in ice, the climate was probably cool.

There are a number of indicators that take us substantially further back in time. They don't do so with the accuracy of the 150-year temperature record. They don't do so with the accuracy of those 1,000-year temperature reconstructions. They don't do so with the accuracy of the hydrogen isotope cores from Antarctica, but they do give us a rough overall picture of what the climate has looked like over the last four billion years, which is about the history of the Earth. The Earth was formed about 4.6 billion years ago. I'm not going to describe the total history in all its gory detail. Again, the more recent times in that history, we know better. But if you look at the history of the Earth's climate going far back, it's probably the case that the present-day is relatively cool compared to the historical climate. By the way, that's somewhat of an interesting anomaly because we know that the Sun has actually been getting brighter over time. That's something that's worried climatologists, but they think they have it figured out.

The present is probably a little bit cooler, and the present has been experiencing this fairly rapid cyclic behavior in the climate associated with the ice ages and the warm interglacials, occurring roughly every 100,000 years. That's a fairly short time scale on the four-billion-year history of the planet Earth. As you go back in time, into times tens to hundreds of millions of years ago—and here we're getting into the age of dinosaurs—it was probably a lot warmer. How much warmer? I don't know. We can't really pin that down, but significantly warmer. There's evidence of tropical-like fossil vegetation found in what, at that time, were continents that were in the Arctic, for example. The Earth was considerably warmer at times in the past, at times tens to hundreds of millions of years ago, during the age of the dinosaurs.

There was a time further back—probably somewhere between 200 million and a billion years ago, maybe—when the Earth's temperature fluctuated quite a bit, but averaged maybe not terribly differently from what it is today. That may extend back to several billion years, maybe two billion years or so, with some considerable fluctuations. Going way back into the early history of the planet, of course, we were formed in a very hot state, the heat being due to the accretion of small bodies in the solar system that congealed together to make planet Earth, and gave up their energy as heat when they collided together. In the very early history of the planet, it was hot. At some point, it was too hot for life. But life formed probably well over three billion years ago, and has been continuing ever since. You have a climate record of a living planet that goes back billions of years.

One of the most interesting things about that record, and something that's only recently come out, is there are probably situations—and they probably occurred somewhere between about 750 and maybe 580 million years ago—in which the entire planet froze solid. We believe these times were associated with periods when continental drift had put most of the continents near the equator. The oceans froze completely solid. The continental interiors were completely dry. Dust blowing off the continents landed on the frozen ice and changed the amount of sunlight that was absorbed. Volcanic eruptions put gases into the atmosphere that, as I'll talk about in subsequent lectures, helped the planet warm. Eventually, the Earth recovered—in fact, probably very quickly—from those states, which are called "snowball Earth."

This is still somewhat controversial. Not all scientists believe they occurred, but there begins to be some pretty good evidence that Earth froze solid on several occasions. If it freezes solid, you might worry, how will it ever melt again? Ice reflects so much sunlight; how could it ever warm up? The answer is there was dust involved, and there were gases from volcanoes involved, that let it warm up again. One of the interesting things about those snowball Earth phases is that we believe that they may have killed off a good fraction of the species that were living on the planet, but right after them, there were vast sort of explosions in the diversity of living things, including possibly the occurrence of land animals and complicated multi-cellular organisms. They may have some of their origin in these unusual climatic events called snowball Earth.

The climate has done interesting things going back in the past. If we were to ask the question we started this lecture with, is the Earth's climate—and particularly the change in climate we've seen in recent decades—unusual? Yes, it's unusual in the context of the last 1,000 years. Is it unusual in the context of Earth's entire three- or four-billion-year history? Absolutely not, at least if you talk about the magnitude of the change. It was a lot cooler than just the degree or so warming we've seen. It was cooler by six times that much during the most recent ice age. During the age of the dinosaurs, it was certainly warmer globally, by a lot, than it is now. The Earth has certainly seen climate extremes, which are much more dramatic than the variation we've seen in the last decade.

The more significant question is, is the variation in the last decade unusually rapid? Have there been very rapid oscillations of the entire global climate? That is a question that needs to be answered. That's a question that's a bit more subtle. We certainly know that, as Earth came out of the most recent ice age, 10,000 years ago or so, there were, at least in regions—parts of the Northern Hemisphere for example—quite rapid excursions in temperature that occurred naturally. Did they occur globally? We're still not sure about that. We don't know the answer to the question of whether the climate change of recent decades is unusually rapid, but it certainly is in the context of the millennium. In the context of global temperature changes, it probably is unusually rapid. As I'm going to show you in subsequent lectures, we understand the cause of the recent climate change, and that cause is things we're doing. That is causing this very rapid climate change of recent decades.

Lecture Four
In the Greenhouse

Scope:

What determines a planet's temperature and, thus, establishes its overall climate? Ultimately, stable climate results from a balance between incoming solar energy and heat that a planet radiates to space. Well-established fundamental principles of physics govern that balance. But a planetary atmosphere complicates the picture, absorbing outgoing energy and, thereby, altering the energy balance. The result is the greenhouse effect, which can keep a planet significantly warmer than it would otherwise be. For Earth, the naturally occurring greenhouse effect—mostly the result of atmospheric water vapor and, to a lesser extent, carbon dioxide— warms our planet by nearly 60°F (33°C) over what it would be absent these gases. Changes in atmospheric composition will alter the quantitative value of this greenhouse effect, and that's the primary reason for concern about human-caused climate change.

Outline

I. *Energy balance* is the key to a stable climate—or a stable temperature in any system, be it a house, a planet, a star, or a pan on a stove.

 A. A greenhouse provides a good example, maintaining a balance between heat loss and incoming sunlight.

 B. A planet works the same way, balancing energy from the Sun (or another star) with infrared energy radiated to space.

  1. We know the rate at which sunlight delivers energy: This averages about 240 watts for every square meter of Earth's surface.

  2. We know the law that governs radiation: The energy loss by radiation is proportional to the fourth power of the temperature.

  3. Equating energy gain and loss gives an average temperature of about −18°C or 0°F.

  4. This is in the right ballpark, but it seems a bit cool for a global average. What's wrong with the calculation?

II. We've neglected Earth's atmosphere, which warms the planet's surface through the *greenhouse effect*.

A. The gases that make up most of the atmosphere—roughly 80% nitrogen and 20% oxygen—are largely transparent both to incoming sunlight and to outgoing infrared.

B. Other gases, especially water vapor and carbon dioxide, are transparent to sunlight but considerably opaque to infrared.

    1. These are called *greenhouse gases* (GHGs). The name is somewhat of a misnomer, because the glass in a greenhouse blocks predominantly the bulk motion of heated air out of the greenhouse, not infrared radiation.

    2. Greenhouse gases tend to have more complicated molecules than nitrogen and oxygen, and that's the fundamental reason for their greater absorption of infrared.

    3. The atmospheric greenhouse gases absorb outgoing infrared radiation, and as a result, Earth's surface warms until it reaches energy balance at a higher temperature than it would have without the greenhouse gases.

    4. The *natural greenhouse effect*, due largely to naturally occurring water vapor but also to naturally occurring carbon dioxide, raises the surface temperature by about 33°C or 60°F. That makes Earth a much more comfortable place for living things.

C. Three ways to understand the greenhouse effect involve increasing levels of scientific sophistication.

    1. Simple: Greenhouse gases act as an insulating "blanket," making it harder for energy to escape and resulting in the warming of the surface until its higher temperature can still force enough energy through to maintain energy balance.

    2. More sophisticated: Greenhouse gases absorb outgoing infrared and heat up. They re-radiate their energy both upward to space and downward to the surface. The surface is warmed by this *back radiation* and, therefore, radiates more infrared to maintain energy balance.

    3. Most sophisticated: The upper atmosphere, being cooler than Earth's surface, can't emit as much radiation as the surface. Radiation from the atmosphere alone, then, can't keep the planet in energy balance. Therefore, the surface temperature must be higher than the atmosphere temperature to provide enough additional radiation that can escape through the nearly but not completely opaque greenhouse gases. Thus, the greenhouse effect depends crucially on the *temperature difference* between surface and upper atmosphere. Increasing greenhouse warming actually entails *cooling* the upper atmosphere.

III. Earth's actual energy balance is more complex still. Other factors include the following. See Figure 7.

    A. Reflection, which returns incoming sunlight to space before it can participate in energy flows that establish surface and atmospheric temperatures.

        1. Reflection occurs from clouds, from ice and snow, from particles suspended in the atmosphere (aerosols), and from deserts and other light-colored surfaces.

        2. Changes in the amount of sunlight reflected will change Earth's energy balance.

    B. Convection, evaporation, and transpiration, which carry energy from the surface into the atmosphere by the bulk motion of heated air and moisture. These processes are affected by surface temperature, vegetation, and other factors.

Suggested Reading:

Houghton, chapter 2.

Going Deeper:

Wolfson, chapter 12.

Harvey, chapter 2 through section 2.3.

Ruddiman, chapter 2.

Web Sites to Visit:

Hadley Centre of the British Meteorological Office, http://www.metoffice.gov.uk; search for brochures. The highly respected British Met (for Meteorological) Office provides a thorough introduction to the greenhouse effect at this page on its Web site.

Questions to Consider:

1. Give explanations of the greenhouse effect at two different levels of scientific sophistication.

2. What is meant by the statement "the natural greenhouse effect amounts to 33°C"? Why is the effect associated with a number?

3. Name two greenhouse gases. Which makes the greatest contribution to the *natural* greenhouse effect?

Lecture Four—Transcript
In the Greenhouse

In the first three lectures, we looked primarily at observational data. We looked at what Earth's temperature had done, first from thermometer measurements going back 150 years. We then looked at other observations of species, of ice sheets, of glaciers, etc., which told us something about what Earth's climate had been doing. We then looked at proxy indicators for pushing climate records back, first hundreds, and then thousands, and then hundreds of thousands, and eventually billions of years. In all those examples, we were looking at observational data. We weren't trying to say what explained the data, why the data did what they did. We were simply observing what had been happening to Earth's climate. We came away with the conclusion that we've had some extremely rapid climate change in the last few decades, that it is unprecedented—certainly in the last 1,000 years—and that its rate may be particularly significant, may be a particularly rapid climate change.

This lecture is very different. Here we look at the big ideas—the big science ideas—behind what establishes a planet's climate. I want to spend a good deal of time on this because I want you to understand thoroughly just how much we know about the basic big idea behind climate. Again, as I said in the first lecture, we don't know every detail, but we have a very good theory of the big picture, and we have very good reasons for believing that theory is correct. In Lecture Four, and then in Lecture Five, I will give you good evidence for why we understand how the climate behaves, and why we understand what establishes climate. In Lecture Five, we'll come back to looking at some observational evidence that these theories are, in fact, correct.

The key idea, the one I want to begin with, is the idea I call *energy balance*. It's the key to a stable climate, and it's also the key to a stable temperature in any system you care to think about. It could be a house; it could be a pan of water setting on the stove. It could be anything that maintains a temperature. It could be your own body, for that matter. What maintains a system at a fixed temperature? The answer is very simple. The answer is basically that energy balance occurs when a house, a planet, a body or anything else, loses heat at a rate that depends on the temperature difference between that thing and its surroundings. That's true of just about everything. If your house is warmer than the surroundings, heat is flowing out through the windows, through the insulation, through the cracks and crannies. Heat is flowing out, and the rate at which the heat flows gets bigger if the temperature difference between inside and outside gets bigger.

The Sun is losing energy because it's very hot, and it radiates that energy off into space, and that radiation, what we call sunlight, is carrying away energy from the Sun, and it's being carried away at exactly the same rate at which energy from nuclear processes in the Sun's interior is supplying energy to the Sun's surface layers. A pan setting on a stove is at the same temperature if the stove burner is supplying energy to it at the same rate at which various processes are sending that energy off into the surrounding environment of the kitchen. The same is true for a planet. A planet is in energy balance, and will maintain a stable temperature if, in fact, the energy loss processes that cause the planet to lose energy are balanced by the energy gain.

The key to understanding energy balance is the idea that the greater the temperature difference between a system and its surroundings, the greater the rate of heat flow. If the system is cooler than its surroundings, it's gaining energy from outside; if the system is warmer than its surroundings, which is usually the case we're talking about, it is losing energy. The rate at which it's losing energy depends on how much hotter it is than its surroundings. That's a big idea to grab hold of because, if that weren't the case, we wouldn't naturally get systems into energy balance. Because that's the case, we do naturally get systems into energy balance—the larger that temperature difference, the larger the rate of heat loss. Therefore, there will come some temperature where the rate of heat loss is equal to the rate at which energy is coming into the system. At that point, we're in balance, and the system will maintain a constant temperature.

This idea of energy balance isn't quite the same as another science idea you may have heard of, called equilibrium. In equilibrium, a system would simply come to the same temperature as its surroundings. That would happen if I took a glass of water and put it here on this table, and just let it warm up to the surroundings. In energy balance, something active is happening. Energy is flowing into a system from outside, and energy is flowing from the system back to the environment. When those two processes are in balance, the system sets at a fixed temperature. I don't care whether it's a house, a body, or a pan on the stove or whatever.

A greenhouse provides a particularly good example of that. If the system—for example, a greenhouse—is hotter than it "ought to be" in its natural state, then it's going to be losing energy at a greater rate than it's gaining energy, and it's going to cool down. If the system is cooler than its surroundings, then it's going to be losing energy at a lesser rate, or energy's going to be coming into it, and there's going to be a net increase in energy, and the temperature is going to go up. It's naturally going to come into a balance in a situation where it's losing energy at the same rate at which it's gaining energy. A greenhouse, again, is a perfectly good example. A greenhouse is simply a building with some glass that lets the sunlight, the energy from the Sun, in.

If the greenhouse is in balance, it's warmer than its surroundings, but the rate at which energy is flowing into the greenhouse is equal to the rate at which energy is flowing out of the greenhouse, and we're in a state of energy balance. If nothing changes—if the Sun doesn't set, if the outside temperature doesn't change—then the greenhouse will set there at whatever temperature it's at, and that temperature will be warmer than the surrounding temperature because it's got a flow of energy coming in, and it's got a flow of energy going out. If the greenhouse is too cool because the rate at which the energy leaves the system depends on the temperature difference between that system and its surroundings, the rate at which energy is leaving the system will be less, but the rate at which it's coming in from sunlight will still be the same. Consequently, there will be a net flow of energy in, and the greenhouse will get warmer; the temperature will rise.

On the other hand, if the greenhouse is hotter than its balance temperature—the temperature it would like to be at in energy balance—then it's warmer than its surroundings, and so the heat flow out is even greater. It's greater than the heat flow in, and as a result, the system is losing energy, and it will cool down. That's the reason why systems—from a greenhouse to a house, to a planet—all come to an energy balance at a fixed temperature that would remain constant if no other circumstances changed. In some systems, like houses, we have active devices like thermostats that control, ultimately, the rate at which the energy is coming into the system from the furnace, and they maintain a temperature that you choose. But in other systems, like the greenhouse or a planet, you don't choose the temperature it's going to be at. The conditions that describe the amount of energy coming in, and that set the amount of energy going out, ultimately determine the temperature.

The point is, either way, there is an energy balance that gets established naturally. If something upsets that balance, the balance will be re-achieved. If conditions change, like you put more insulation in the greenhouse, or you do something to Earth's atmosphere, then the balance condition will change, but you will eventually achieve a new balance. As long as the conditions don't keep changing, that balance gets achieved naturally. Again, the fundamental science reason why it gets achieved is because the rate at which the system loses energy depends on the temperature difference between that system and its surroundings. The details depend on the mechanism whereby that energy leaves the system. But the fact is that the bigger the temperature difference between a system and its surroundings, the bigger the rate of heat loss, and that's what allows a system to come naturally into balance at some temperature.

How does this work for a planet? In the case of Earth, for example, the energy input to planet Earth is the incoming sunlight. We know the rate at which sunlight comes to Earth. Averaged over the entire planet, averaged over night and day, taking account of sunlight that bounces off ice sheets and never gets to participate in the Earth's energy system, sunlight comes in at the rate of about 240 watts on every square meter of surface area. For every square meter, about a square yard of Earth's surface, energy is reaching the Earth's system from the Sun at the rate of about 240 watts; that is, on each of those square meters. What's a watt? I'll give you more about that much later in the course, but you know what 100-watt light bulb is, so 240 watts is about 2.5—not quite 2.5—100-watt light bulbs worth of energy arriving on every square meter of Earth from the Sun.

If Earth is to be in energy balance, it has to lose that energy. The Earth, being an isolated system in empty space, has only one way to lose energy. It radiates it by what's called electromagnetic radiation. Because Earth's temperature is what it is, around 15°C, around 60°F, around 300 Kelvin—that's degrees Celsius above absolute zero—it happens that the energy it radiates is in the form of infrared radiation that we can't see. It's radiation with a longer wavelength than the visible light that we see with our eyes, and it's radiation that is invisible infrared. That's how the Earth loses its energy, through infrared radiation.

The Sun, being much hotter than the Earth—6,000 Kelvin, about 6,000°C, about 11,000°F—loses energy also by electromagnetic radiation because it's surround by the vacuum of outer space; but because it's so much hotter, its light is visible. That distinction is really important. The Sun is providing us with energy in the form of visible light. The planet Earth is re-radiating energy in the form of infrared radiation. They're both forms of electromagnetic waves—I talk about those in some detail in my *Physics in Your Life* course, what electromagnetic waves are—but they're different. The incoming sunlight is visible to us with our eyes; it's very short wavelength radiation. The outgoing infrared, because the Earth is so much cooler than the Sun, is infrared. That's going to play a major role in climate.

Before we get to that role, let's just talk about the energy balance that gets achieved. If you had an Earth that was initially very cold, it would be losing no energy by infrared radiation. Sunlight would be coming in. The planet would be gaining energy; it would be warming up. As it warmed up, it would radiate more and more infrared radiation. We know very well the physical law that describes the rate at which a hot object radiates energy. In fact, it goes as the fourth power of the temperature, T^4. Therefore, as the temperature increases—that's absolute temperature, measured from absolute zero—the amount of energy radiated goes up very rapidly. As our initially cold Earth begins to warm up, it radiates more and more infrared. Eventually, when it's radiating an equivalent of 240 watts for every square meter of surface, it will be radiating, totally, exactly as much energy as it's gained from the Sun. It will come into energy balance, and the Earth will therefore be at a fixed temperature. That's how Earth's energy balance gets established.

If for some reason the Earth were to get hotter, it would be radiating at a greater rate than that 240 watts/m^2 coming in from the Sun, and therefore it would cool down. If it were a little cooler, it would be radiating at less than the rate at which it was gaining energy from the Sun, and it would then warm up and eventually come back into energy balance, as long as nothing changed. Nothing meaning the sunlight's intensity didn't change, and the properties of the Earth, and particularly its atmosphere that allows this radiation to escape, didn't change. As long as none of that changed, we would have a stable climate. Again, I'm kind of talking here about an average condition. Surely things vary from place. They vary from night to day. They vary with the seasons. I'm kind of ignoring all that and talking about a sort of steady long-term average climate.

We can do better than just talk qualitatively about Earth's energy balance because we know the law, again, that tells the rate at which a system emits electromagnetic radiation. The system emits electromagnetic radiation at a rate that depends on its temperature and, as I said, on the fourth power of the temperature. If we equate the equation—and I'm not going to do the math here—the term that gives the rate at which a hot object radiates energy to 240 watts on every square meter, the rate at which sunlight is coming in to Earth, we'll have an equation that has one unknown, the temperature, and we can solve for it. When we do, we find that the Earth, by this simple energy balance argument, ought to have an average temperature of about -18°C, which works out to about 0°F.

Think about that a minute. It's not way off because we could have had a temperature that was so hot that the oceans had boiled away, but it's not that hot. It could have been so cold that the entire planet was frozen in a snowball state. Well, if it's 0°F on average, probably there are at least some places that are going to be warmer by quite a bit; enough, probably, that water is a liquid. On the other hand, an average Earth temperature of 0°F or -18°C sounds a little bit low, and it is. So what's wrong with our calculation? What's wrong with our calculation is that we've neglected the Earth's atmosphere. Here we have to get into the details, the crucial details that determine why Earth's climate is warmer, in fact, than it would be if the Earth's atmosphere weren't there.

Most of the atmosphere, as you probably know, is composed of nitrogen, N_2, which is basically inert and plays very little role in chemical processes; a little bit, but not much. About 20% of it is oxygen, which is crucial to our life, and to the life of most organisms. Both oxygen and nitrogen, which incidentally consist of simple diatomic molecules, O_2 and N_2, are largely transparent to this incoming visible radiation from the Sun. A little bit of it is absorbed in the atmosphere, but most of it comes right through to the surface.

They're transparent to that visible. They're also transparent to the outgoing infrared radiation. However, there are other substances in the atmosphere, trace gases, gases that are there in only small amounts that have a significant effect on the ability of infrared radiation to escape the atmosphere. These gases are largely transparent, again, to the incoming visible sunlight that's bringing energy to the Earth, but they're largely opaque to the outgoing infrared.

What are those gases? In the natural Earth system, the dominant ones that are most important are water vapor, which depending on the humidity, may be a few percent of the atmosphere; and CO_2, which is considerably less than a percent of the atmosphere, but is nevertheless very significant in terms of its influence on the outgoing infrared. In the natural system, the natural state of things, water vapor is the most significant of these gases. These gases are called greenhouse gases. That name is a bit of a misnomer. The idea is that they function sort of like the glass in a greenhouse, but they don't really. The glass in a greenhouse keeps heated air from escaping the greenhouse in bulk; these gases keep radiation from escaping. That's a bit different than keeping heated air from escaping. The glass of a greenhouse does have some effect in keeping infrared from escaping, but that's a minor effect. Nevertheless, this term greenhouse gas has stuck, and this effect, whereby greenhouse gases trap outgoing infrared radiation, is called the greenhouse effect.

If you want to know a little bit more of the physics of why the greenhouse effect occurs—I'm not going to go into the details. Maybe a chemist would call this the chemistry of it; I'd call it the physics. If you look at the two molecules I just mentioned, water (H_2O) and carbon dioxide (CO_2), those are both triatomic molecules, or molecules that have three different atoms. It turns out those triatomic molecules have many more ways to vibrate, rotate and move around than the simple diatomic molecules of oxygen and nitrogen. Many of those ways involve vibrations at the frequencies of infrared radiation. That's why water vapor and CO_2 are greenhouse gases. They tend to be more complicated molecules than the simple molecules, oxygen and nitrogen, that make up most of the atmosphere.

These atmospheric greenhouse gases absorb outgoing infrared radiation. They act like insulation. They keep that infrared radiation from getting out; they kind of trap it. I'll describe this in a little more detail in just a moment. The net effect, though, is that Earth's surface has to be at a higher temperature to get rid of that 240 watts/m^2 of radiation, of energy coming in. It has to be at a higher temperature than it would be otherwise. In the case of the natural greenhouse effect caused by naturally occurring water vapor and CO_2 in the Earth's atmosphere, the value of that effect, the amount by which it increases Earth's surface temperature, is about 33°C, roughly 59 or 60°F. The Earth is warmer than that number we calculated, that 0°F, by about 60°F. It's warmer than that -18°C by about 33°C. That gives it a temperature, on the average, of about 15°C or 60°F.

The Earth is warmer and more comfortable than it would otherwise be because of this naturally occurring greenhouse effect. When I talk about the natural greenhouse effect, I'm talking about the effect of CO_2 and water vapor in absorbing outgoing infrared—water vapor has the biggest effect; CO_2 comes second—and therefore warming the surface to levels above what it would otherwise have. When I say the natural greenhouse effect is 33°, I'm talking about something very specific. It's 33°C warming due to the presence of these greenhouse gases. I think you'll agree that a planet at 60°F average temperature, about 15°C average temperature, is a lot more comfortable than one that would have been at 0°F, -18°C. It wouldn't have been an uninhabitable planet. I think life might have still formed, and some forms of life could still exist in some areas if there weren't the greenhouse effect, but it sure makes for a more habitable, enjoyable planet.

That's the essence of the greenhouse effect, and I'd like to spend a little bit more time describing not only the essence of the greenhouse effect, but also three different ways of understanding it in terms of scientific understanding. The simplest way is just to say that greenhouse gases act as an insulating blanket. They make it harder for energy to escape, and therefore the surface temperature is a little higher than it would be otherwise. I think every citizen of an industrialized country that produces greenhouse gases ought to know at least that. A slightly more sophisticated way of looking at that is to consider an Earth system with greenhouse gases in the atmosphere. In an Earth system with greenhouse gases in the atmosphere, the infrared going out gets absorbed in the atmosphere, some of it. It heats up the atmosphere.

The atmosphere warms, and it does what any other warm system does; that is, it radiates energy out in the form of infrared in all directions. For the atmosphere, all directions means upward and downward. The atmosphere returns to the surface some of the radiation that it got back from the surface originally. Consequently, the temperature of the Earth's surface has to be warmer than it would be otherwise, in order for that energy to escape and keep the planet in energy balance. If you had an Earth without greenhouse gases, and you added greenhouse gases, that would momentarily drop the rate at which the Earth lost energy by infrared radiation. It would therefore heat up until there was again a balanced established, at what would be then a higher temperature.

Perhaps the most sophisticated way of looking at all this is to consider what happens with the upper atmosphere. Because the upper atmosphere is a cooler place than the surface and the lower atmosphere, the upper atmosphere can't radiate as much energy as the surface can because it's cooler. In particular, it can't radiate away all the energy that comes in as sunlight. Therefore, the Earth's surface has to get warmer and warmer in order to get some of that energy to escape, despite the presence of the greenhouse gases. If for some reason the upper atmosphere were at the same temperature as the surface, there would be no greenhouse effect. That's an important point that I hinted at in earlier lectures, in Lecture Two when I talked about the temperature change that's occurred recently in the stratosphere. The stratosphere has actually cooled. A cool upper atmosphere and a warm surface are conditions needed for the greenhouse effect to occur.

That's the most sophisticated understanding of the greenhouse effect. If you were to sort of diagram that, you would imagine that a little bit of infrared is escaping to space directly from the atmosphere—or a significant amount, maybe—but not enough to balance the incoming sunlight. You need some more additional infrared to escape from the surface. Because of the greenhouse gases, you need to have the surface very warm because it has to get a significant amount of energy out through the nearly opaque greenhouse gases. You can begin to see that there are many ways that you might upset this kind of balance. You may upset this kind of balance by changing the amount of greenhouse gases in the atmosphere. You might upset this balance by changing the amount of sunlight. You might affect this balance by changing the composition of the greenhouse gases. There

©2007 The Teaching Company

are many ways to upset this balance, but the point is it's a delicate balance, and it's established by a balance between the incoming sunlight and the outgoing infrared.

Earth's energy balance is actually a lot more complicated that what I've just described here. I want to show a diagram that gives you a sense of that complexity. I've only talked about, overall, in comes this much sunlight, out goes this much infrared. There's actually a lot of complicated interactions occurring, particularly between Earth's surface and its atmosphere. To understand climate in a little more detail, we need to understand those interactions. I'm going to take a look at a diagram that describes some of the details of Earth's energy balance. Again, this is still averaged over the entire planet. We're not talking about what happens in Greenland, in the tropics, in Hawaii, or in the middle of the South Pacific Ocean. We're not talking about that. We're just looking at overall global averages. Already, it begins to be fairly complicated and sophisticated.

I was recently, by the way, at a climate conference that I go to every year, a conference that relates to the Sun and its influence on climate. Every single speaker showed this diagram. By the time the third or fourth speaker had shown it, that became kind of a standing joke. Here is a diagram that depicts pretty much our best understanding of what the energy balance looks like in all its detail. Different scientists will give you slightly different values for some of the numbers that appear in this diagram—that's a matter of ongoing research—but the big picture, we sort of understand. In this diagram, there is a very large arrow representing the incoming solar radiation. It's about 342 watts on every square meter. Wait a minute; I thought you said 240. Well, I did, but 342 watts/m^2 come in; some of those watts are reflected back into space before they ever get to play a role in the climate system. They're just sunlight that's reflected off into space. What are they reflected from? From the tops of clouds, which are highly reflective. If you've flown in an airplane and looked down on the clouds, they're very white and bright. That means sunlight is being reflected back into space.

If it goes back into space reflected, it never gets absorbed by the Earth's atmosphere system, and it never plays a role in climate; except, in a sense, a negative role by not being there to provide energy for the Earth's atmosphere system. Of that 342, about 77 of them are reflected by clouds and by particles and things in the atmosphere. About 30 of those watts are reflected by the surface, largely by ice sheets, snow, deserts, and light-colored areas. There's another way we could upset Earth's energy balance, by changing the amount of snow there is on the ground, or by deforesting an area and turning it into a desert. That would tend to make it absorb less sunlight, and more sunlight would be reflected. That radiation, a total of 107 watts/m^2, going off on the upper left of the diagram, doesn't play a role in the climate system. It doesn't get absorbed.

Then what happens to the rest of the energy that's coming in, which amounts to about 235 watts/m^2? That's my 240, roughly, that I was talking about before. A small amount of it, about 67 watts/m^2, gets absorbed in the atmosphere in clouds, by volcanic dust, by dust that is blown off deserts, by particulate air pollution. They get absorbed by things in the atmosphere and deposit their energy as heat in the atmosphere. Roughly only half of the incident solar energy—here it's 168 watts/m^2—actually gets absorbed and warms the surface directly. Some of it's already been reflected; some of it got stopped in the atmosphere. Roughly half of it gets to the surface, and its job there is to warm the surface.

Absent any atmosphere, that 168 watts/m^2—but of course, we've already had an atmosphere to reflect some of that. But if there were no other effects in the atmosphere, that would be sent off as infrared radiation to space, and the Earth's surface temperature would be whatever it needed to be to radiate that 168 watts/m^2. But we have the greenhouse gases in the atmosphere. On the right side of the diagram, you see the effect of the greenhouse gases. You see that the greenhouse gases are radiating back to the surface; 324, in this particular diagram, watts/m^2 back to the surface. Of course, everything is in balance, so the surface is radiating most of that back away again. Most of it gets stuck in the atmosphere.

There's almost a cycling, if you want to think of it that way. It's not building up; everything's in a steady balance. But there's a small fraction of that, depicted by an arrow that goes all the way from the surface up to the top of the atmosphere, about 40 watts/m^2; it's called the atmospheric window. That's the small amount of energy that's able to escape from the surface, all the way up to space, as infrared radiation because the greenhouses gases are not 100% opaque. If they were 100 % opaque, we'd be in big trouble, but they're not, and that atmospheric window allows a small amount of energy to escape.

It's enough to keep everything in balance. The point is, because of those greenhouse gases radiating energy back downward, the Earth's surface has to radiate a lot more energy upward. Because the radiation increases with temperature, that means that temperature has to be a lot higher than it would be otherwise. How much higher? That's 33°C, about 60°F, of the natural greenhouse effect.

That's the picture. If we had to look at this diagram and say where's the greenhouse effect, it's in that upward and downward—particularly in that downward flowing arrow, representing back radiation that's slowing down from the greenhouse gases in the atmosphere to the surface. Of course, those greenhouse gases are all sending some radiation off into space. This isn't the whole picture. The reason it isn't the whole picture is because there are other important processes that transport energy between Earth and atmosphere. In particular, they consist of convection, a process whereby heated air gets less dense and rises, and it takes with it the energy that it contains by virtue of being hotter. That's, in this diagram, called thermals. When you see a day when there are a lot of little puffy cumulus clouds forming in the atmosphere, they typically form at the top of thermals. Warm air has heated; it's risen, and it's cooled a bit as it rises. When it reaches the point where the moisture condenses, you get a cloud at the top of that rising thermal. What that thermal has done is transported energy into the atmosphere. A significant flow of energy occurs because of those thermals; the actual bulk transport of heated air into the atmosphere.

A more significant flow from the surface into the atmosphere is in what's called latent heat. That's the energy that's associated with water being in the gaseous state instead of the liquid state. When water evaporates from the surface of the ocean, and that moisture-laden air is transported into the upper atmosphere, it's carrying with it that moisture. It carries with that moisture, in the form of vaporous water, the energy that it took to turn liquid water into water vapor. If that water vapor recondenses into liquid water and falls as precipitation, that energy becomes available as heat. It's that energy that, for example, in the extreme case, drives hurricanes. It's largely the energy associated with the evaporation of water. Of course, that was ultimately solar energy, but in this case, the mechanism for transferring the energy into the atmosphere is not radiation—not the sending out of infrared radiation because things are hot—but rather the transfer of moist air containing the so-called latent heat associated with the evaporated water.

The energy balance of Earth is a complicated system, and we need to understand the details of that system to understand climate. The big picture for us here in this lecture is that energy balance will occur naturally, and it will occur when the rate at which energy comes into a system, particularly the Earth, is equal to the rate at which energy goes out from the system. For Earth, the energy goes out entirely to space by infrared radiation, but there are transfers between the Earth and the atmosphere; some of them by radiation, some of them by these convective effects, the thermals and the latent heat, that move energy back and forth between Earth and the atmosphere. The bit important one for our purpose is the effect of the greenhouse gases in absorbing the infrared, sending some of them back to the surface, and making the Earth warmer than it would be otherwise by about 33°C or 60°F, the natural greenhouse effect.

Lecture Five
A Tale of Three Planets

Scope:

We can't do controlled experiments to verify the theory of the greenhouse effect. However, we can look at past climates to see how temperatures and greenhouse gases are correlated. Also, nature provides us with an "experiment" in the three planets Venus, Earth, and Mars. Mars's atmosphere is so thin that it has almost no greenhouse effect. Venus's dense atmosphere results in a runaway greenhouse effect that increases the planet's temperature by hundreds of degrees. Earth lies in between, with its 33°C greenhouse effect.

But isn't this still "just a theory"? It is—but in science, *theory* doesn't mean hypothesis or speculation. It means a coherent body of scientific understanding, with a conceptual framework supported by diverse evidence. The theory of the greenhouse effect is one such body of understanding. That doesn't mean we understand every detail of Earth's climate, but it does mean that our basic understanding is grounded in solid scientific principles.

Outline

I. How do we know that greenhouse gases are associated with the warming of Earth's surface? One way is to look at past climates and their relation to greenhouse-gas concentrations.

   A. Although water vapor is the dominant greenhouse gas for Earth's natural greenhouse effect, carbon dioxide (CO_2) plays a more important role in varying the strength of the greenhouse effect.

      1. That's because water-vapor concentration adjusts nearly instantaneously to changing conditions, while added CO_2 remains in the atmosphere for a long time (more on this in Lecture Six).

      2. The same ice cores that provide a record of past temperatures also give scientists a record of atmospheric composition. These data come from analysis of ancient air bubbles trapped in the ice. For so-called *trace gases*—those other than nitrogen and oxygen—

concentrations are measured in parts per million (ppm) or parts per billion (ppb) by volume.

  3. Temperature and CO_2 concentration show a remarkable correlation. See Figure 8.

B. There's not a simple cause-and-effect relation here. Rather, subtle changes in Earth's orbital parameters result in a slight warming; that, in turn, causes CO_2 release to the atmosphere. More CO_2 means more greenhouse warming and more CO_2 release. The temperature "leapfrogs" upward through such *feedback mechanisms*.

C. A similar relation between temperature and CO_2 concentration holds farther back. During the age of dinosaurs, around 100 million years ago, warm-climate organisms thrived even in polar regions. CO_2 levels at this time probably exceeded 1000 ppm. Geologic processes— plate tectonics, volcanism, and rock weathering—are the major causes of climate change on these long time scales.

II. Nature also provides us with a climate "experiment" in the form of our neighbor planets Venus and Mars.

  A. A simple calculation based on energy balance with no atmosphere yields average temperatures of:

  1. −50°C for Mars.
  2. −18°C for Earth (as described in Lecture Four).
  3. 55°C for Venus.

  B. Actual surface temperatures are about the same for Mars (−63 K), a bit warmer for Earth (15°C), and a lot warmer for Venus (about 500°C—hot enough to melt lead). Why the discrepancies?

  1. Earth's 33°C difference is due to the greenhouse effect, as described in the preceding lecture.
  2. Mars's atmosphere has only about 1% the density of Earth's, so it has a negligible greenhouse effect. It's actually cooler because of reflection of sunlight.
  3. Venus's atmosphere has 100 times the density of Earth's, and it's about 96% carbon dioxide. Although Venus's proximity to the Sun would make it only modestly warmer than Earth, that was enough to evaporate Venus's surface water, resulting in a

"runaway" greenhouse effect powered at first by water vapor and later maintained by carbon dioxide.

C. Our neighbor planets confirm the theory of the greenhouse effect: that greenhouse gases warm a planet's surface beyond the temperature that would result from energy balance in the absence of those gases.

D. Earth appears safe from a Venus-like runaway greenhouse effect, but Earth is still subject to warming if the concentration of greenhouse gases increases.

III. Isn't this all "just a theory" that might or might not prove correct?

A. "Just a theory" is a misleading phrase. In science, a theory is a set of coherent principles that explain a broad range of physical phenomena. A theory gains credibility when it's supported by evidence from observations and laboratory experiments, and it's rejected when physical reality presents evidence that contradicts the theory. Established theories have survived that test. Some examples of well-established scientific theories include:

1. The theory of relativity in physics, which describes phenomena ranging from the overall structure of the Universe, to numerous astrophysical events, to the functioning of the global positioning system.

2. The theory of evolution in biology, which describes the origin of species through the evolutionary process.

3. The theory of plate tectonics in geology, which describes the movement of the continents over geologic time.

4. The theory of the greenhouse effect, which describes the fundamental principles that establish planetary climates.

B. A theory need not have explanations for every detail in order to qualify as being a solidly established body of scientific knowledge. Although evolution doesn't yet describe every step in the origin of every species, that doesn't invalidate the theory of evolution. Although relativity can't tell us everything that goes on inside a black hole, that doesn't invalidate the theory of relativity. And although climate scientists don't understand every detail of Earth's complex climate system, that doesn't invalidate the theory of the greenhouse effect.

Suggested Reading:

Houghton, chapter 4, p. 66 onward, emphasizing CO_2-temperature connection.

Wolfson, chapter 12, section 5.

Going Deeper:

Ruddiman, chapters 4 and 8.

Schneider, Rosencranz, and Niles, chapter 1, especially p. 42 onward.

Web Sites to Visit:

Carbon Dioxide Information Analysis Center, http://cdiac.ornl.gov/trends/ co2/contents.htm. At this particular CDIAC page, you'll find graphs and data on atmospheric carbon dioxide from sites around the world and on time scales ranging from decades to hundreds of thousands of years.

Questions to Consider:

1. Graphs of temperature and carbon dioxide concentration over the past 160,000 years show a remarkable correlation. Why can't we draw the simple conclusion that the warm periods result from an increase in carbon dioxide, leading to an increase in temperature?

2. Venus receives only twice as much energy from the Sun as does Earth, and most of that is reflected back into space. Yet Venus is much hotter than Earth. Why?

3. What's a scientific theory? How is it more than just a hypothesis or a guess?

Lecture Five—Transcript
A Tale of Three Planets

Lecture Five: "A Tale of Three Planets"—with apologies to Charles Dickens, I suppose. We've seen evidence that Earth's climate is changing. Then in the most recent lecture, we saw the theory of what establishes a planet's climate. That theory involving the basic idea of energy balance, that the energy that arrives from the planet's star, in the case of our planet—and in fact, our neighboring planets that I'll be discussing in this lecture—energy arriving from the Sun; that that energy heats up the planet, and the planet sheds the energy back to space as infrared radiation. Regardless of whatever else goes on, we know that if the sunlight is shining steadily and the conditions of the planet are steady, the planet will come to an energy balance at a constant temperature, and that temperature will be maintained, essentially unchanged, as long as none of the conditions—the incoming sunlight or the ability of the planet to shed that energy back to space—change. On the other hand, if any of those conditions change, the planet will come to a new energy balance at, perhaps, a new temperature.

We've seen Earth's climate changing. We've come to an understanding of how climate arises in planets, but I've given you basically there a theoretical understanding. How do we know that that's correct? How do we know that this theory that energy balance establishes a planetary climate and particularly the idea of the theory of the greenhouse effect—the idea that there are gases in the atmosphere of a planet, particularly of Earth that are opaque to outgoing infrared radiation, and therefore they require a surface temperature that's warmer than it would be in the absence of those gases. How do we know these theories are true? One of the nice things scientists like to do is experiment with nature. They like to do controlled experiments. They say, "Let's set up this condition, and let's put more greenhouse gases in the atmosphere, and let's see what happens to climate." They do carefully controlled experiments.

Unfortunately, Earth's climate is one area where you can't easily do carefully controlled experiments. You can't play with the climate. We can't decide we're going to double the amount of greenhouse gases in the atmosphere. We're going to see what happens. We're going to decide if we like what happens, and then we're going to undo the experiment. We can't do that. We don't have multiple

Earths to work with, to perform different experiments on. It would take a long time for climate to change, even if we could do experiments. It's difficult to establish miniature climate experiments. It's impossible to put a tiny Earth in the laboratory and do realistic climate experiments.

Having said all that, that doesn't mean there aren't some things we can know experimentally. For example, we can put carbon dioxide (CO_2) in a chamber. We can send infrared radiation through it. We can measure how much of the infrared radiation gets through. Consequently, we can determine the opacity; that is, how much the CO_2 blocks infrared radiation. We can then determine, if we put so much CO_2 in the atmosphere, how much infrared radiation will it block. That's something we can measure experimentally, for example. As I'll describe in a subsequent lecture, we can also build, if you will, simulated climate systems—Earths, and atmospheres, and oceans, and Suns, and all the rest—and put them in computer models whose mathematical programming represents the physical, biological, and chemical science going into these systems. We can run these models and try to understand how climate would behave under other circumstances.

Climate scientists who do that actually call these computer runs experiments. But they aren't experiments in the sense of our grabbing onto nature itself and doing things and trying to figure out what happens. We can't do controlled experiments with Earth's climate, unfortunately. On the other hand, there are many who would argue we are, in fact, doing inadvertent experiments on Earth's climate by in fact, altering the amount of greenhouse gases in the planet's atmosphere. But those aren't controlled experiments. Those aren't experiments we are intentionally doing, and they aren't experiments where we're controlling exactly the effect we're producing.

We can't do experiments to verify the theory of the greenhouse effect that I described for you in the last lecture, so how do we know it's correct? Well, I'm going to give you, in this lecture, several arguments why we are quite certain that the greenhouse effect is a serious and important explanation of what is happening to our planet's climate, what establishes the climate originally, and what's happening now as the climate is changing. One argument I would give you is that the basic physics I described in the previous lecture—the physics that goes into the greenhouse effect; the absorption of infrared radiation by CO_2; the presence of CO_2 and water vapor, both absorbing infrared radiation in the atmosphere— these are things we can get in the laboratory. We know, for example, that law that I referred to last time about the amount of infrared radiation depending on the fourth power of the temperature; that's something we can verify in the laboratory.

In other words, that whole theory of the greenhouse effect that I described to you in the last lecture is something that's based in very sound basic scientific principles that we know from theory, and we know from measurement, to be true. It's not some cooked-up theory designed to explain one particular thing—namely in the Earth's climate. It's a theory based on very fundamental, very sound, very long-established scientific principles. So there's one reason why the theory I described to you last time—and I'm using that word theory intentionally, and at the end of this lecture I'm going to talk a little bit about what scientists mean specifically when they say "a theory." One reason we believe the greenhouse effect theory to be correct is because it's based in very sound and very fundamental science.

What's another way? In my arguments last time, I suggested that the presence of the greenhouse effect—the 33°C warming that the Earth experiences relative to what it would have if there were no greenhouse gases in the atmosphere—that that is caused by the presence of water vapor and CO_2 in the atmosphere. Perhaps, if we look at the atmospheric CO_2 content and the atmospheric water vapor content over time, and we looked at Earth's temperature over time, and we saw a correlation, we might be able to argue that, in fact, the presence of one infers the climate change associated with the other, for example. I'm going to try to do that in the first part of this lecture, but I want to warn you at first that this is a little bit more subtle than a very quick glimpse at the data might suggest.

First of all, water vapor, I argued last time, is the predominant greenhouse gas in the Earth's atmosphere in terms of the natural greenhouse effect. That's true; people tend to forget that. When they hear greenhouse effect, they tend to think CO_2. But in fact, the natural greenhouse effect is dominated by water vapor. CO_2 is important and significant, but it's not the dominant greenhouse gas. CO_2, though, plays a much greater role in regulating the strength of the greenhouse gases and the strength of the greenhouse effect. The reason water vapor is less important in that regard is because water vapor adjusts itself in the atmosphere almost instantaneously to the conditions in the atmosphere.

If the atmosphere were to warm, for example, causing more evaporation, and therefore more water in the atmosphere, that effect would be felt almost immediately. The time scales associated with changes in the water content of the atmosphere are on the order of a week. What happens to the water? It evaporates; gets into the atmosphere, within about a week, it precipitates back out and lands on the surface. So, the water vapor content of the atmosphere adjusts very, very quickly to changing climatic conditions.

Not so CO_2; particularly not so CO_2. I'm going to present you some reasons in the next lecture about precisely why that is, and have you come to an understanding of why CO_2 is so unique. But for the present, the important point is this. Changes in the CO_2 content of the atmosphere are most significant in causing changes in climate, we think. I'm going to show you why precisely we think that. Although water vapor is the most important natural greenhouse gas, CO_2 plays a much bigger role in regulating climate.

What we'd like to know about is what the CO_2 content of the atmosphere has looked like over time. How are we going to measure that? If we want to go substantially far back in time, those very same ice cores that gave us these isotopic temperature measurements—the measurements of temperature based on the ratio of oxygen or hydrogen isotopes; the heavier isotopes being more slow to evaporate and more soon to precipitate out that giving us an indicating of temperature at the time the precipitation fell that has now become ice—we can more easily measure the CO_2 content of the atmosphere at that same time, because trapped in that ice are small air bubbles. They contain, if you will, fossil air from the time that that ice was formed.

Now there are some subtleties here, as always in science. For example, that air diffuses somewhat through the ice, and there are corrections that have to be applied for that. Scientists know about these corrections, and they believe they're applying them correctly. If you want to get precise dating of the CO_2 bubbles, you have to apply a little bit of a correction. We know about that; we understand that. When we do that, we can get a record of past atmospheric composition; in particular, past CO_2 levels. How do we measure

©2007 The Teaching Company

that? We measure it in concentration, typically in a measure called parts per million, ppm; sometimes for even rarer gases, in parts per billion (ppb). Strictly speaking, you have to specify something more besides saying ppm or ppb. You say ppm by volume, or ppm by weight, or whatever.

When we're talking about atmospheric gases, it's essentially always ppm by volume, which you will sometimes see abbreviated ppmv, or just ppm. If you want a picture, what that means is, imagine a milk jug—one-gallon milk jug—and imagine you took a million of those, and you filled them up with air. If you had done this in 1750, before the industrial revolution started and before humankind started burning fossil fuels in abundance, you would have found that, out of those million milk jugs, each a gallon in size, each containing air, if you had separated out the atmospheric constituents in the pre-industrial era, about 280 of those milk jugs would have contained CO_2. That's what it means when I say the pre-industrial concentration of CO_2 was about 280 ppm. Of all the volume in the air, by volume, about 280 gallons out of every million gallons was CO_2. That number has changed, and we'll talk about that change in number subsequently, in the lectures where I talk about the human influence on climate. Until humankind began putting a lot of CO_2 into the atmosphere, the amount we had, the concentration, was about 280 ppm, and that's what that means.

What we would like to do now is consider those same ice cores that I showed you data from earlier—and I showed you at that point only temperature—and we're going to look now at both temperature and CO_2. I want to show a graph again; this is a graph you've seen before—horizontal axis is the time in thousands of years before the present. I'm going to show you one cycle of that repeating, roughly 100,000-year period; that's the ice age cycle. We've seen this graph before. It shows the deviation in temperature from the present, which is read as zero on the vertical axis here, the deviation in degrees Celsius. You'll remember there was a number I wanted you to walk away from this course with, and that number was the rough difference between the present and sort of the middle era of an ice age. That was on the order of 6° or 8°C; you can quibble about the numbers. That sort of sets the scale for the temperature variation.

You saw that characteristic pattern that repeated again over about 500,000 years, of a fairly steep rise into these interglacial periods of moderate temperatures, the kind of temperatures the Earth is experiencing now, followed by a long decline into this cool ice age period. There's a lot of fluctuations within that ice age. It's not a smooth decline; it's not a steady cold temperature. Lot of fluctuations, and then something triggers this rapid warming. We've seen that before, and that's what this picture shows. What if we put on top of that same graph the CO_2 concentration as measured from the same ice cores? Instead of working here with oxygen isotope ratios or hydrogen isotope ratios, we're working instead with actual analysis of the air trapped in these ancient air bubbles, the fossil air that's in there. We can measure the CO_2 concentration.

The most remarkable thing about the two graphs—the upper line being the CO_2, whose scale is at the right; the lower one being the temperature whose scale is at the left—the most remarkable thing is they seem to track almost perfectly together. They have exactly the same shape. They have that 100,000-year periodicity between highs in temperature and highs in CO_2. They have that long decline into lower temperatures and lower CO_2. There is a remarkable correlation between the Earth's average temperature—at least to the extent that these Arctic and Antarctic ice cores reflect Earth's average temperature, and we think they do—there's a remarkable correlation between CO_2 concentration and global temperature. With a lot of other data sets going back a lot further in time, we have very good reason to believe that correlation holds to a very good extent, throughout much of Earth's history, if not all of it. There is a very strong and obvious correlation between temperature and CO_2 concentration.

Before you read too much into this, it is not the case here that somehow the Earth's CO_2 content increased, the atmosphere's CO_2 increased and the temperature went up with it, and then the CO_2 went down, and the temperature went down with it. It's not as simple as that. As I suggested when I first showed you the temperature graph a few lectures back, there's a subtle cause and effect thing going on here that involves orbital changes in the parameters associated with the Earth's orbit. Some of those changes include the eccentricity of the orbit, how elliptical it is. Earth's orbit is not a perfect circle; it's slightly elliptical. Earth is about 3% closer to the Sun at its closest approach than it is at its most distant point. If I were to ask you to think a second, when is it closest to the Sun? You might want to pause your DVD a second, or pause your tape, and listen. Think about that. I'll answer it for you. The Earth happens to be closest to the Sun in January. That's when the Northern Hemisphere has its winter, and that effect makes the winter/summer difference less substantial than it would be.

If that were to change, and it changes partly with changes in the eccentricity, the ellipticity of the orbit; it changes partly with changes in the tilt of the Earth's axis at 23 degrees. That change occurs on about a 41,000-year cycle. The orbital eccentricity changes on about 100,000- and a 400,000-year cycle. There's also a change associated with the so-called precession of the equinoxes, the fact that the North Pole doesn't always point in the same direction; but on a cycle of about 25,000 years, precesses around. I showed you why in my video series *Physics in Your Life*. It's like a big top precessing; it takes 25,000 years. So the North Star is now in the direction of the Earth's polar axis, and 12,500 years from now it won't be; and 25,000 years from now it will be again. All those effects conspire to make subtle changes in the climate.

For example, at some point, the closest point to the Sun will be in June. When that happens, the summer/winter difference will be exaggerated. You might say well, doesn't that just reverse in the Southern Hemisphere? It doesn't because the Southern Hemisphere has a lot less land, a lot more water, and so the effects are different. There are all these subtle effects. What we believe happens to trigger this simultaneous change in CO_2 and temperature is there are these orbital effects. They're small, but they're regular; they're periodic. They're small effects on the Earth's orbit, but they're big effects on the climate because they trigger a small change in temperature. Then we believe other changes occur.

For example, the oceans get a little warmer. That means some of the CO_2 that's dissolved in the oceans comes out, goes into the atmosphere, and makes more greenhouse warming that makes more CO_2 come out of the atmosphere or the soils. There are a host of other changes like this called positive feedback effects that we believe leapfrog and bring about that fairly rapid rise in both temperature and CO_2, until we reach these interglacial temperature maxima, like we're in right now. Then gradually, as the orbital parameters change a little bit, those conditions go away. Gradually again, working in step, the CO_2 goes down, and the temperature goes down, and that drives the CO_2 down. They go down step by step. It isn't quite "Yes, CO_2 increases, and so does temperature." In the long history of the Earth, they go up and down together. The important point is they're strongly correlated.

That's evidence that this greenhouse effect is really happening, that the CO_2 content of the atmosphere helps determine what the temperature of the atmosphere is; never mind the fact that the CO_2 and the temperature kind of worked on each other to get up and down in that 100,000-year cycle we've been talking about. This relationship, we're pretty sure, holds a lot further back. For example, during the age of the dinosaurs—which was on the order of 100 million years ago, 60 million years ago, 200 million years ago, that kind of time—we know that warm-climate organisms existed in what are now tropical regions. CO_2 levels then were maybe 1,000 ppm, maybe about four times what they were before the start of the industrial era, and the temperatures were far higher than anything the planet is looking at now.

Over the very longest term in the history of the Earth, it's not so much orbital processes, like I've described here for the ice ages, but geological processes—volcanism, volcanoes putting gases into the atmosphere; plate tectonics, the movement of the continents to different places on the Earth; and the weathering of rocks, which can release carbon into the atmosphere or take carbon out of the atmosphere. Those are believed to be the main drivers of climate over the time scales of hundreds of millions of years and billions of years. We believe, and have good evidence, that the CO_2 content of the atmosphere was high during times when the temperature was also high. One of our big pieces of evidence that supports the theory of the greenhouse effect is this very tight correlation between CO_2 and temperature over many, many time scales in the history of the Earth's atmosphere and climate.

But still it would be nice to do an experiment, but we can't, as I argued at the beginning of this lecture. But nature has done an experiment for us, and it's provided us with our neighbor planets, Venus and Mars. You'll recall that I kind of outlined a simple calculation in which I said the Earth receives energy at approximately the rate of 240 watts/m^2 approximately, and it must therefore get rid of energy at the rate of 240 watts/m^2. We know the rate at which a hot thing gets rid of energy by infrared radiation, and that depends on its temperature. That sets us up an equation that we can solve for temperature.

We can do the same for the other planets. This gets a little bit subtle, but if you assume those other planets are basically just chunks of rock with no atmosphere, and you do the same calculation. Venus is closer to the Sun, so you would expect it to be warmer. It turns out that, because of that energy loss going as T^4, you don't have to change the temperature very much to compensate for Venus's proximity to the Sun. If you calculate what Venus's temperature ought to be, given that very simple calculation I sort of did for Earth, or outlined for you for Earth a few lectures ago, you get that Venus ought to be at about 55°C. That doesn't sound real hot. It's hotter than it is here on Earth. It would not be comfortable for us. It's nowhere near the boiling point of water, on the other hand, which is 100°C. It's different from Earth, but not dramatically different.

You do the same calculation for Mars, you get something on the order of about -50°C. Now that's cold, but it's not so cold that it would freeze the atmosphere or something, like Pluto, whose atmosphere is frozen on the rocks of the planet's surface. Now it's cold, but it's not that cold. That's because Mars is about 50% further from the Sun than the Earth is. That's why Venus is a bit closer, about two-thirds as far out as Earth is. That would explain these relatively minor differences in the temperatures of these three planets. However, as I say, nature has provided us with an experiment in the form of the planets Venus, Earth and Mars. Earth, of course, if you do that calculation—and we already did that, basically—and we found that Earth ought to be at about -18°C, 0°F. Here's the numbers again; -50°C for Mars; -18°C for Earth; and getting closer to the Sun, about 55°C for Venus.

What are the actual surface temperatures? We know those well. We have spacecraft roving around the surface of Mars. I showed you in the first lecture how we calculate the average surface temperature for the Earth. We've had spacecraft go through the Venutian atmosphere. We know pretty well the composition of these planets, of their atmospheres, and so on; and we know what the temperatures, the surface temperatures, are on these planets. If you look at Venus, Earth and Mars, and look at the calculated temperatures—55°C for Venus, -18°C for Earth, -50°C for Mars—and you ask what's the actual temperature on these planets? There's a dramatically different answer for each one. For Venus, it's especially dramatic. Venus's actual surface temperature is around 500°C, far above the boiling point of water, above the melting point of lead. This is a hot place. This calculation is way off for Venus, in a way that it wasn't way off for Earth.

What's going on? You do the calculation for Earth, we already know that the actual temperature is 33°C, 60°F, warmer than it would be if we didn't account for the greenhouse effect. That's why the Earth's average temperature is 15°C. There's a 33°C-greenhouse effect for Earth. Mars's actual temperature is actually pretty close to the calculated temperature. It's certainly the closest of all of them. It's actually a little bit cooler, and there are some subtle effects going on here. I have included in these calculations the effect of reflection of sunlight from Earth and Mars in doing these calculations. I haven't included it for Venus. Venus, it turns out, is so cloudy that it reflects nearly all the sunlight coming onto it. In fact, if you did a naïve calculation that said somehow Venus reflects almost all the sunlight that comes onto it, but it doesn't have any atmosphere—an impossibility, but if you did that calculation, you would get a calculated temperature for Venus that's actually less than Earth's temperature because Venus reflects so much of its sunlight. But that's not what's happening.

What is going on here? We already know what's going on for Earth. We know that Earth has about a 33°C-natural greenhouse effect. That's the effect of the greenhouse gases in Earth's atmosphere blocking the outgoing infrared or, more subtly, absorbing the outgoing infrared, re-radiating it to the surface, and then the surface has to be warmer to get rid of that energy. What's going on on Mars? Mars has an atmosphere that is about 1% the density of Earth's. It's a very tenuous atmosphere; it's hardly any atmosphere at all, and it exerts essentially no greenhouse effect. So, Mars's climate is not significantly altered by a natural greenhouse effect on Mars. But you look at Venus. Venus's atmosphere is 100 times the density of Earth's—a huge amount of gas there—and it's 96% CO_2. Earth's, before humankind began fiddling with it, was about 280 ppm. Venus', if I said it in ppm, would be 960,000 ppm—96% CO_2, a very powerful greenhouse gas.

Absent that, Venus's proximity to the Sun would make it only mildly warmer than Earth. But it was enough mildly warmer that a lot of

Venus's early surface water, we believe, evaporated, forming a water vapor-based strong greenhouse effect. That caused a cascade of events that eventually resulted in a lot of CO_2-based greenhouse effect in this strong CO_2 atmosphere. Venus has what we call a runaway greenhouse effect. The greenhouse effect went wild; it built on itself through positive feedback mechanisms, and brought Venus up to this impossibly inhospitable 500°C-temperature it now has. Venus and Earth are two very similar planets. They're the same size; they're same mass, approximately the same distance from their star. But that distance difference is enough to cause this runaway greenhouse effect to occur on Venus, and Venus now has this inhospitable temperature due to its 96% CO_2 atmosphere. We are pretty sure that a runaway greenhouse effect is impossible on Earth, so don't worry about the oceans boiling away here on Earth. That doesn't mean we shouldn't worry about greenhouse warming on Earth, but we are unlikely to go into that runaway state that Venus is in.

I would argue that our neighboring planets provide this natural experiment that confirms for us that the theory of the greenhouse effect is not some kluged together explanation to tell us what's happening on Earth, but it's kind of a universal principle that talks about planets that have atmospheres, and describes the climates of those planets in terms of the composition of their atmospheres, and particularly the role that CO_2 plays in establishing the natural greenhouse effect in those planets. That's the theory of the greenhouse effect. I introduced that in the last lecture, and this is the confirmation for the theory of the greenhouse effect; the basing on firm scientific principles, the correlation between Earth's temperature and its CO_2 concentration; and finally the tale of three planets, the greenhouse effects being very different on the three planets—Mars having almost none, Earth having a modest 33°C-greenhouse effect, and Venus heating up by hundreds of degrees Celsius because of its greenhouse effect.

Let me pause and get a little bit here into philosophy of science. Because isn't all of this just a theory? That phrase "just a theory" is being batted about a lot these days to talk about scientific theories ranging from greenhouse warming to evolution. I want to say a little bit about the phrase "just a theory." I think it's a misleading phrase, "just a theory." What's a theory? In science, a theory is not just some hypothesis or vague idea someone puts out and says maybe this is true. A theory in science is a coherent body of principles—principles, big underlying principles—that describes some body of scientific knowledge; and particularly, describes many different instances. For example, the theory of the greenhouse effect describing what's going on on three very different planets; that's an example of explaining a broad range of scientific phenomena. These theories gain credibility as they're supported by evidence from observations, by laboratory experiments; and theories are rejected if physical reality conspires to disprove the theory.

Established theories—theories that have been around for a long time—have survived many, many, many tests. Every bit of scientific evidence has to corroborate the theory, or at least not contradict it. If any bit comes along that contradicts it, and can be repeatedly verified to contradict it, then the theory has to be replaced. There are many established theories that have survived these tests. Let me just give you a few examples. The theory of relativity in physics, which I describe in my course *Einstein's Relativity and the Quantum Revolution*, my Teaching Company course; *Modern Physics for Non-Scientists* is the subtitle. That deals with phenomena ranging from the overall structure of the Universe to astrophysical events, to the functioning of the global positioning system. If we didn't understand Einstein's theory of relativity, we wouldn't be able to explain any of those diverse phenomena, from the origin of the Universe to the behavior of the global positioning system satellites. That's a remarkably solid—established theory. No reputable scientist doubts that relativity is part of a correct description of physical reality. That's a long-established theory.

In biology, the theory of evolution, which describes how species evolved through the evolutionary process, is a very well established, solid theory. In geology—and this is one of the newcomers to this list of big theories—the theory of plate tectonics; the idea that the continents move around over the surface of the Earth over time, driven by convective motions in the Earth's mantle. That's a relative newcomer, but that has formed a body of theory that then coherently describes a great many phenomena that we observe in the structure of our planet. I would add to that set of big theories—not quite as big as evolution, or relativity, or maybe even plate tectonics, but nevertheless as well-established and based on scientific principles, and explaining a broad range of observed phenomena—the theory of greenhouse effect that describes the fundamental principles that establish planetary climates

I want to emphasize one other thing I said in the first lecture. A theory doesn't have to have explanations for absolutely everything in order to qualify as a solidly established body of evidence. Evolution doesn't describe every step in the evolution of every species. Plate tectonics doesn't describe absolutely everything that happened. We still don't know what goes on inside black holes because relativity doesn't answer that for us. We still don't understand every aspect of the climate system; what's the role of the carbon that's coming out of the soils or methane from plants? We don't understand all that. It doesn't matter in terms of the coherence and validity of the big picture idea.

Climate scientists don't understand every detail of the climate system. But to point to a detail we don't understand is not to invalidate the theory of the greenhouse effect, any more than to say I don't know exactly how this monkey species evolved, and I don't know exactly what's going on in this black hole, invalidates the long established theories of relativity or evolution. There we are, with a solidly established understanding of what brings about planetary climates. An essential key to that is the well-established and scientifically-understood greenhouse effect.

Lecture Six
Global Recycling

Scope:

The water vapor and carbon dioxide that create Earth's greenhouse effect don't remain in the atmosphere forever. They cycle through the atmosphere, oceans, soil and rocks, and living things. A typical water molecule remains in the atmosphere about a week; a carbon dioxide molecule, about five years. Plants remove carbon dioxide from the atmosphere and produce energy-storing foods through photosynthesis. Plants and animals use this energy, taking in oxygen and returning carbon dioxide to the atmosphere. There's also carbon exchange between atmosphere and oceans. And a tiny fraction of the carbon gets stored as fossil fuels. Carbon dioxide emissions from human combustion of fossil fuels constitute a small fraction of the global carbon cycle—but that carbon accumulates in the system, pushing up the atmospheric concentration. It will take centuries to millennia for this extra anthropogenic carbon to leave the system.

Outline

I. With respect to matter, Earth is essentially a closed system. The materials that make up the planet and its atmosphere and oceans never disappear, but some cycle continually among different parts of the Earth system.

    A. This contrasts with energy. With respect to energy, Earth isn't closed; energy arrives as sunlight and returns to space as infrared radiation. The preceding lectures described this energy flow, and Lecture Ten will explore it further.

    B. Many material cycles are important to climate, including the water cycle and the carbon, nitrogen, phosphorus, and sulfur cycles. But of these, water and carbon are the most important, in part because both water vapor and carbon dioxide are significant greenhouse gases.

II. The *water cycle*, or *hydrologic cycle*, involves the evaporation of surface water and its subsequent precipitation onto land and oceans.

    A. Solar energy ultimately drives the water cycle, with the rate of evaporation strongly dependent on temperature.

B. The amount of water in the atmosphere is tiny compared with the amount stored in the oceans and cryosphere (frozen water). Although it varies with time and location, water vapor typically constitutes a few percent of the lower atmosphere.

C. Atmospheric water vapor responds quickly to changing temperature and other conditions.

    1. A typical water molecule remains in the atmosphere for only about a week before it's removed by precipitation.

    2. From the long-term perspective of climate, therefore, atmospheric water vapor can be considered to adjust almost instantaneously to changing surface conditions.

    3. Atmospheric water vapor contains the energy—called *latent heat*—that was required to evaporate it. Therefore, the water cycle also contributes to the flow of energy from surface to atmosphere, as discussed in Lecture Four.

III. The *carbon cycle* operates on many time scales and involves interactions among atmosphere, land, oceans, biological systems, and on the longest time scales, geological processes. See Figure 9.

    A. Let's first define the terms *tons, tonnes,* and *gigatonnes*.

        1. In the United States, 1 ton is 2000 pounds. In the rest of the world, 1 ton means 1 *metric ton*, usually spelled *tonne*. One tonne is 1000 kilograms, or 2200 pounds. Thus, a U.S. ton and a metric ton are about the same. Numbers given here are in metric tons, but the difference isn't significant.

        2. The metric system uses prefixes for powers of 10, such as kilometer (km) for 1000 meters and kilogram (kg) for 1000 grams. The prefix *giga* (G) stands for billion; thus, 1 *gigatonne* (Gt) is a billion tonnes. That's the unit used for global carbon flows. A gigatonne is the same as a *petagram* (1000 trillion grams, or 10^{15} grams).

    B. The most rapid part of the carbon cycle involves carbon exchanges between air and water, air and living things, and air and soil.

1. Atmospheric carbon dioxide dissolves in water, a process that removes CO_2 from the air. But dissolved CO_2 also escapes to the atmosphere. Normally, these two processes are nearly in balance, and the flows each way amount to about 90 Gt of carbon (measured as C, not CO_2) per year. Because warmer water holds less CO_2, warming of the ocean surface sends more CO_2 into the atmosphere.

2. Plants take CO_2 from the atmosphere through *photosynthesis*, using solar energy to form carbohydrates that are the fuels for living things. Photosynthesis removes about 110 Gt of carbon from the atmosphere each year. In the process, plants release oxygen to the atmosphere.

3. Plants and animals use the energy stored in carbohydrates for cellular activities and locomotion. This process of *respiration* takes oxygen from the atmosphere and returns CO_2. Respiration, along with the decay of dead and waste matter, returns nearly all of the 110 Gt of CO_2 that was removed in photosynthesis. Rivers carry some additional carbon into the oceans.

4. The atmosphere, living things, soils, and surface ocean waters all represent short-term carbon *reservoirs*. Cycling among these reservoirs occurs mostly on relatively short time scales. In particular, a typical carbon dioxide molecule remains in the atmosphere only about five years.

5. But the rapid cycling of carbon through the atmosphere–biosphere–surface ocean system means that any carbon added to that system remains there much longer—for hundreds to thousands of years. Because the added carbon cycles through the atmosphere, the level of atmospheric carbon dioxide goes up and stays up for a long time.

C. Two processes are involved in the long-term removal of carbon from the rapidly cycling atmosphere–biosphere–surface ocean system.

   1. Marine organisms die and sink into the deep ocean, taking their carbon with them. This is the *biological pump*.

2. Upwelling of deep water and upward diffusion of carbon nearly balance the biological pump.
3. The result is a net downward flow of carbon of only about 2 Gt per year. That's why it takes hundreds to thousands of years to remove excess carbon from the system.
4. This carbon joins the huge (39,000-Gt) reservoir of carbon stored in the deep ocean.
5. Carbon slowly leaves the deep ocean reservoir and is incorporated into ocean sediments, which eventually form rocks.

D. Two additional processes operate on very long time scales of tens to hundreds of millions of years.
1. Geological processes eventually recycle buried carbon through volcanism.
2. A tiny fraction of carbon fixed by photosynthesis is buried before it has a chance to decay. Over millions of years, this becomes the fossil fuels coal, oil, and natural gas. We're burning those fuels now, adding CO_2 to the atmosphere.

IV. Let's briefly review the first half of the course.
A. Earth is warming, with especially rapid warming evident in recent decades of the 150-year thermometric temperature record.
B. A host of other evidence corroborates the thermometric record.
C. Using proxy indicators pushes the climate record back further in time.
1. Studies suggest that the warming of the late 20th and early 21st centuries is unprecedented in the past 1000 years.
2. Ice-core records show a pattern of cold ice ages punctuated by shorter, warmer interglacial spells lasting a few tens of thousands of years. For the past half million years, this pattern has repeated on a roughly 100,000-year time scale.

D. We understand the fundamental processes that establish planetary climates based on solid physical principles.

1. Stable climate entails a balance between incoming sunlight and outgoing infrared radiation.
2. Infrared-absorbing greenhouse gases in a planet's atmosphere alter the details of this balance, causing the planet's surface to warm. For Earth, the most important greenhouse gases are water vapor and carbon dioxide. Together, they provide a natural greenhouse warming of about 33°C or 60°F.
3. Ice-core data show that atmospheric carbon dioxide and temperature have been closely correlated for nearly a million years, and less direct evidence suggests that such a correlation continues back through Earth's history.
4. Neighbor planets Mars and Venus are strikingly different, and their differences help confirm the theory of the greenhouse effect.

E. Cycling of materials plays a role in climate, with the most important cycles being those of water and carbon.
   1. The carbon cycle involves rapid cycling of carbon between atmospheric carbon dioxide and the biosphere, soils, and ocean surface waters.
   2. Carbon added to this system stays there for centuries to millennia and adds to the atmospheric carbon content.

Suggested Reading:

Houghton, chapter 3 through p. 39.

Wolfson, chapter 13, section 5.

Going Deeper:

IPCC 4, chapter 7, first two sections.

Harvey, chapter 2, sections 3–4.

Web Sites to Visit:

Hadley Centre, http://www.metoffice.gov.uk. Thorough discussion of the carbon cycle.

Questions to Consider:

1. What is the biological pump, and how does it help remove carbon from the Earth–atmosphere–surface ocean system?

2. A typical molecule of carbon dioxide remains in the atmosphere for only about five years. Why, then, does the addition of CO_2 cause the atmospheric carbon dioxide concentration to remain increased for hundreds to thousands of years?

Lecture Six—Transcript
Global Recycling

Lecture Six: "Global Recycling." We've established that greenhouse gases—substances that absorb infrared radiation going out from the surface of the planet—play a major role in establishing a planet's climate. The greenhouse gases that I particularly emphasized as being important for the natural greenhouse effect are water vapor and CO_2. We want to ask a little bit more about what these substances do in their natural behavior in the planetary system because, as you already know for water, they don't just set there in the atmosphere. Water comes out as precipitation; it's re-evaporated; it flows through rivers; it goes into underground water aquifers, and so on. What do these substances do? We really have some sense of the water cycle already, but what about carbon, particularly? This lecture is going to emphasize the role of carbon, the cycling of carbon through the Earth's system and, to a lesser extent, water because we already understand most of what we need to know about water.

But CO_2, and carbon in particular, has some very subtle aspects to its behavior that are really important to understand if you want to grasp why it is that we humans have the capability we do to interfere with the climate system, because you might argue that we just play a very small role in all this, and the numbers suggest that we play a pretty small role. How can we end up making a major impact on climate? I want to give you some detailed background science, particularly on the behavior of carbon that will help you to understand this. I want to begin, though, with an important distinction with respect to matter. Earth is basically a closed system.

What I mean by that is that any, say, oxygen that the Earth possesses, it's probably possessed since its origin. That oxygen may be assuming different forms. Sometimes it's part of water, H_2O; sometimes it's O_2 in the atmosphere; sometimes it's incorporated into living things, in the carbohydrates we eat as foods. It may move around and change its chemical state, but it's still the same oxygen the Earth has had all the time. The Earth has iron in its core and throughout its lithosphere, its rocky material. Some of that iron gets incorporated into the hemoglobin of our blood, and then returns to the natural systems when we die and decay. That's iron cycling around through the natural system. Most materials do that; they undergo natural cycles. But with respect to these materials, the Earth is a closed system.

I want to contrast that with energy, which has formed been an important player in the first part of the course, and is going to become an important player in the last part. With respect to energy, Earth is not a closed system. Energy comes into Earth, almost all of it from the Sun—as we'll see in Lecture Ten, not quite all of it, but almost all of it comes from the Sun. It goes back out into space in the form of infrared radiation. To the extent that we are in energy balance—and we are very, very nearly in energy balance, even if not perfect—the extent that we're in energy balance, all the energy that comes to us from the Sun is radiated back into space. The Earth processes energy as sort of a once-through system. The Earth is not a closed system with respect to energy.

But with respect to matter, with the exception of a few molecules that occasionally escape off the upper atmosphere into space, and some meteoritic dust, which comes down through the atmosphere, and the occasional gas from a comet tail that we pass through, or the occasional spacecraft that we send out into the cosmos, never to return to Earth—with those minor exceptions, the materials that make up our planet and its atmosphere, and its oceans, and its rocks, and so on, cycle continuously through the Earth atmosphere/ocean system. We need to understand those cycles because it's when some materials, particularly water and CO_2, are in the atmosphere that they exert a greenhouse effect. When they're not in the atmosphere, they don't. So, we need to understand these cycles.

There are lots of cycles that are important, as I've suggested. In particular for climate, the water cycle and the carbon cycle are important. The nitrogen cycle turns out to be important—I mentioned nitrogen as an atmospheric constituent that is largely inert—but it does play an important role, for example, in fertilizing plants. There are plants that can take nitrogen out of the atmosphere and incorporate it into plant matter, and that makes protein eventually. Lightning strikes can fix nitrogen from the atmosphere into other nitrogen compounds, so there is some cycling of nitrogen through the system. Phosphorus, another nutrient, cycles through the system; sulfur; all these things have some effect when they get into the atmosphere, and so they affect climate. But the important ones are water vapor and CO_2. So, those are the only ones I'm going to talk about. That's because they are present in fairly substantial amounts in the atmosphere compared to many of the other trace substances, and they are strong greenhouse gases. Although what I mean by strong, I'm going to have to explain a little more in a subsequent lecture.

You really know already pretty much what you need to know about the water cycle or the hydrologic cycle. It's basically the cycle whereby water evaporates from the surface of oceans and lakes, rises into the atmosphere, carrying with it that latent heat that I described earlier, maybe moves to a different geographical location, falls as precipitation, typically flows down through the system of rivers and returns to the ocean. It really is a cycle. That's the water cycle. It's solar energy that ultimately drives that cycle, providing the energy that evaporates the water, making molecules move faster, and therefore able to escape the liquid surface and go into the atmosphere as water vapor. That's what drives it. That makes the rate of evaporation very strongly dependent on temperature; in fact, more strongly dependent than you might think. The rate of evaporation goes up very dramatically with temperature until you reach the boiling point, and evaporation then occurs extraordinarily rapidly as the water turns almost immediately into steam.

There's still very little water in the atmosphere. I've mentioned that, depending on the humidity, the water in the atmosphere may be as much as a few percent, or it may be a fraction of a percent, but that amount is tiny compared to the water that's stored in the oceans and in the cryosphere; that is, the body of frozen water. The liquid water, surface waters, underground waters, and the waters locked up in ice constitute a far greater fraction of the water. It's only a tiny amount of the total water on Earth that's in the atmosphere. As I mentioned before, it responds extremely quickly to changing atmospheric conditions. If it gets a little warmer, we get more evaporation. If it gets a little cooler, we get less evaporation, and the water vapor in the atmosphere responds.

One way of thinking about that is to ask this question. How long would a typical water molecule remain in the atmosphere? That's called the residence time. The answer for a typical water molecule is a time on the order of a week. If a water molecule gets evaporated from the surface waters, it finds itself in the atmosphere, and it's likely, on average, to remain there for a week. That's an average.

Some water molecule may come out a day later; some may be up there for a month. But on the average, the time scale for water vapor to live in the atmosphere or stay in the atmosphere is about a week. It's up there, and it has not only its water-ness; it also has this latent energy associated with the fact that it's in the atmosphere as a vapor, and it can give up that energy if it condenses. It may still be in the atmosphere as a cloud droplet, but eventually it precipitates out as rain or snow.

The water cycle does two things. It contributes to a greenhouse gas that's in the atmosphere, but it cycles rapidly, so the amount of that greenhouse gas can change very rapidly with temperature. I'll have more to say about that in subsequent lectures when I talk about the human-induced greenhouse effect. It also contributes to the flow of energy between Earth's surface and atmosphere. That's the water cycle, and you've all seen pictures of the water cycle with mountains, and the water falling down and running down to the rivers. That's pretty well known. The important point I want to emphasize about it is that the water cycle is pretty fast. Water cycles out of the atmosphere on time scales on the order of about a week. By the way, some of the other cycles I mentioned—the phosphorus cycle, the sulfur cycle, and so on—they also have characteristic time scales for the time that, say, sulfur compounds remain in the atmosphere. It's pretty easy to give an unambiguous answer for some of those substances.

Carbon is different. Carbon cycles on many, many different time scales. If someone asks how long does carbon live in the atmosphere, that's a rather subtle question. I want to spend the rest of this lecture exploring the carbon cycle, so we can come to understand the answer to that question. In the back of my mind is something that's been implicit in everything I've said so far, but I haven't' really talked about explicitly. That is, if we change the CO_2 content of the atmosphere—if we increase it in particular—we will strengthen the greenhouse effect, and we will cause the planet to warm. That isn't describing any details, but it's describing the basic physics of what we should expect happens if we put more CO_2 in the atmosphere.

Therefore, it becomes crucial to understand, if we *do* put more CO_2 in the atmosphere, how long is it going to stay there? How quickly can we turn it off? If we cause the planet to warm up, and we don't like that, how quickly can we reverse the situation? If the carbon comes out of the atmosphere in a week like water vapor does, no problem. If it comes out in a year or five years, maybe no problem. But if it takes 1,000 years, may be a problem. We want to explore that. The answer, I'm going to tell you right off the bat, is ambiguous in a subtle, important way that you need to understand to understand why it is we worry about anthropogenic carbon.

The carbon cycle is complicated because it involves interactions among a whole lot of regions. The atmosphere interacts directly with the land surface, with the soils and rocks. The atmosphere interacts with the surface of the ocean, particularly through carbon dissolving in the ocean and coming back out of the ocean. Most importantly, perhaps, the carbon in the atmosphere interacts with the biosphere— that is, with the living content of the Earth—by being taken up by plants and being breathed out by plants and animals as they respire. We want to look at all these processes. There are even geological processes on the longest time scales where carbon in the atmosphere interacts with carbon in the Earth.

The rest of this lecture is going to focus on the carbon cycle, and we're going to look at one diagram that really sums up the entire carbon cycle. It's a diagram that basically shows amounts of carbon stored in different regions of the Earth's system. For example, the atmosphere has stored carbon in it. These places that store carbon are called reservoirs. The atmosphere is a reservoir of carbon. The biota—the plants and animals—are a reservoir of carbon. The surface waters of the ocean are a reservoir of carbon. The deep ocean is a different reservoir of carbon, different for reasons I'll describe. Among these reservoirs are flows of carbon that go back and forth. So, to understand the question how long will carbon stay in the atmosphere, we have to understand all these flows that take place between the atmosphere and the biosphere, and the atmosphere and the soils, and the atmosphere and the ocean, and the upper ocean, and the lower ocean, and so on.

I will say right at the outset that neither the numbers I'm showing in this diagram, nor all the details of the carbon cycle, are fully understood. A former student of mine is one of the experts on this, and she will tell you that there really is no answer to some of these questions yet. Scientists are still activity studying parts of the carbon cycle. But we know enough again, to get a pretty big picture, and to get it pretty quantitatively accurate. Here's what the picture looks like. The picture looks like these reservoirs—the atmosphere, the surface oceans, the deep oceans, the biota, and so on—and they contain certain amounts of carbon. Now, I have to give you an aside here on the measurement systems I'm going to use in discussing this carbon cycle. I'm going to discuss in this lecture only the natural carbon cycle. In a subsequent lecture, I'm going to show you how we humans have altered the carbon cycle, so you can understand that. But everything I'm saying here is about the carbon cycle as it existed before human beings began modifying it, basically by burning fossil fuels, which produce CO_2, which goes into the atmosphere.

But before I do that, a word about measurement systems: The units in this picture are either in gigatonnes (Gt) if they're a reservoir with an amount of carbon; or Gt/year if it's a flow. We've got two things going on here. We've got reservoirs that contain amounts of carbon, and we've got flows that involve carbon moving from one reservoir to another. The amounts are in Gt, and the flows are in Gt/year. What's a Gt, and what's it a Gt of? "Giga" is a prefix that means a billion. You probably, 10 years ago, wouldn't have known that when your computer speeds were measured in megahertz; but today your computer speed, if you've got a good computer, is certainly measured in gigahertz—billions of cycles of operations per second; giga means a billion. It's the opposite of nano, which means a billionth. Everybody is abuzz about nanotechnology, technology at the billionth of a meter scale. Giga is the opposite; it's a billion. Giga means a billion; a Gt is a billion tons.

What's a ton? Well, that's also a little bit subtle. In the United States, a ton is 2,000 pounds, and it's spelled T-O-N. In the rest of the world, a ton is a metric ton, a ton in the metric system, the system international of units, the SI system. In the metric system, a ton is a nice round number, 1,000 kilograms. But a kilogram is about 2.2 pounds. Therefore, a metric ton is about 2,200 pounds. So, a metric ton—2,200 pounds, 1,000 kg—and an U.S. type ton, about 2,000 pounds, are about the same. They're different by 10%, but that's small enough I'm not going to worry about that difference. From now on when I say ton, you're welcome to think English-type ton, the kind of tons we use in the American system of units, if you like, and you're welcome to think metric ton. It doesn't matter because, to within 10%, they're basically the same. I'm talking about Gt mostly;

they're the units used to measure global carbon flows and global carbon reservoir quantities.

You may hear other scientists, especially biologists, talking not so much about Gt, but petagrams. You may wonder what on Earth the relation between those two is. "Peta" is a prefix in the metric system that means 10^{15}. So, it turns out that a petagram, 10^{15} grams, is the same as a Gt, which is a billion—that is, 10^9—kilograms, and a kilogram is 1,000 grams. Anyway, a Gt and a petagram are basically the same thing because a Gt is a billion tons; a billion tons is a billion thousand kilograms, and a kilogram is 1,000 grams. It all works out. A Gt and a petagram are the same thing, if you happen to hear those units. I'm going to talk in Gt here, and then I'm going to ask the last question, Gt of what? The answer is Gt of carbon, not of CO_2. That's an important thing to get straight.

Whenever you see someone saying we Americans are responsible for x-many tons of CO_2 every year, you have to ask yourself, are we talking about tons of carbon or tons of CO_2? Because carbon, when it's tied up as CO_2, has those two extra oxygens with it, and a CO_2 molecule weighs—let's see, I'd have to figure this out. The atomic weight of oxygen is 16, so I've got two oxygens; that's 32. The atomic weight of carbon is 12; that's 44—it weighs 44 atomic units, roughly, and carbon weighs 12, so a CO_2 molecule weighs 44/12 times as much as a carbon molecule. If I have carbon as CO_2, the amount of CO_2 I have is 44/12 or point something times as much as the amount of carbon I've got.

However, in the carbon cycle, carbon is not staying as CO_2; it may be taking different forms. In your body, it's in the form of complicated organic compounds. In the soils, it's in all kinds of forms. In the rocks, it's weathered into different things. It's calcium carbonate in the shells of marine organisms. It takes different forms, and so it makes sense to talk about the carbon cycle not as being about CO_2, but as being about carbon, which sometimes, but not always, takes the form of CO_2. In my diagram, the reservoir amounts are amounts of carbon in Gt, billions of tons, a metric or American; doesn't matter—the flows are in Gt/year of carbon. If you want to convert, for instance, the atmospheric carbon content of 560 Gt—and remember again, that's natural, pre-industrial—you want to convert to CO_2 because it's mostly, but not all, in the form of CO_2, you'd multiply by 44/12, and you'd get the amount of CO_2 in Gt. That's what this picture is all about. Let's take a look at it in detail.

First of all, there's a rapid part of the carbon cycle that involves the exchanges between air and water, between air and living things, and between air and soil. One of the easiest exchanges to think about is what happens when you open a bottle of soda, and the carbon comes fizzing out into the atmosphere; or you leave a bottle of soda, and it gets warm, and even more carbon dioxide comes out, and it's lost its fizz. Carbon dioxide dissolves in water. The warmer the water, the easier it is for it to dissolve. So, there is a constant flow of CO_2 from the atmosphere dissolving in the surface waters of the ocean. However, there's a nearly equal flow in the opposite direction. In my diagram, these are represented by a couple of arrows, upward and downward, representing the exchange of CO_2 between the air and the surface waters. You'll notice the numbers associated with those flows are almost identical. There some slight imbalances, here, but they're all compensated for if you add up all these numbers.

That is the natural carbon cycle. There is one of the easiest parts to understand, the exchange of carbon dioxide between the surface waters and the atmosphere. That's roughly in balance, within a few Gt. What are some other aspects of that? Well, plants play a major role in the carbon cycle. Plants take CO_2 out of the atmosphere, and they use CO_2, which you can think of as burnt carbon; it's combined with oxygen. They basically use solar energy to basically un-burn that carbon, separating the carbon from oxygen, combining with hydrogen from water, and making the so-called carbohydrates, the simple sugars and other carbon, hydrogen, oxygen compounds, that we use as fuels, and that living things in general use a fuels.

This diagram shows that photosynthesis removes about 110 Gt a year of carbon from the atmosphere each year. In the process, the plants release oxygen to the atmosphere, and the original photosynthetic plants are the source of the oxygen that's in the atmosphere. Earth is an unusual planet, to have a strongly oxygen atmosphere because oxygen is very reactive chemically, and it won't stay there for long unless something is replenishing it. What's replenishing it are the green plants, through the process of photosynthesis. I want to focus a minute on that number, 110 Gt/year; that's a big number. It's a lot bigger than the amount of carbon that we humans are putting in the atmosphere from burning fossil fuels, much bigger. But for reasons

I'm trying to show you in this lecture, the amount we put in is still significant. I'll come to that in the next lecture.

What do plants and animals do with this energy? They use it for cellular activity. For big animals like ourselves that can move around, they use it for actual motion. The energy we get all comes from "burning" those carbohydrates, undergoing chemical reactions that convert the carbon and hydrogen that are tied up in the carbohydrates to CO_2 and water. We exude the CO_2 and water; that's why your car steams up when it's cold on the windows and you're breathing out. You're breathing out water. Where did the water come from? It came from the oxygen you breathed in combining with the hydrogen in the hydrocarbons. There's also CO_2 building up in the car. That came from the carbon that was in the fuels.

This process of respiration returns some of the carbon from photosynthesis back to the atmosphere. Something like roughly half of it goes back, for that reason, from the plants themselves; and roughly the other half goes back from the processes of respiration in animals; which have eaten the plants, and also from decay of waste material that we excrete, or from dead organisms decaying. Together, all those processes—respiration, waste material, and decay of dead organisms—return basically all the carbon, not quite all, to the atmosphere, and back it goes. The carbon cycle is almost closed right there in that process.

Some of the carbon ends up getting stored in soils, the surface soils, associated with the presence of living things or dead matter, so there's a reservoir of carbon in the soils also; a reservoir of carbon in the plants and in the animals; not as much in the animals as in the plants. Very, very, very crudely, the sizes of that those reservoirs, the biospheric reservoir and the soil reservoir, are roughly the same order of magnitude as the size of the atmospheric reservoir. The plants and animals are maybe on the order of 500 Gt, maybe a factor of two different. I'm not sure; the biologists aren't absolutely sure. We could study the Earth enough to get that number, but somewhere in that ballpark. There's probably more in the soil, maybe a couple thousand Gt, but it's still roughly that same order of magnitude. We understand about those flows.

We could ask a question—there's a couple other things I should mention here in this rapidly-cycling part of the carbon cycle. The slight imbalance in carbon associated with the cycling between the atmosphere and the biosphere system; a very small amount of carbon ends up in the rivers, and is carried into the oceans. That sort of couples the ocean and atmosphere part of the carbon cycle with the biosphere and atmosphere part of the carbon cycle, which includes both marine organisms and terrestrial organisms. The question we might ask is: How long does a typical carbon molecule stay in the atmosphere as CO_2 in the context of what I've just been discussing, the rapid recycling; carbon removed from the atmosphere by photosynthesis, carbon goes back from respiration, carbon comes out of the oceans when they get warmer, the carbon goes back into the oceans. How long does an average CO_2 molecule stay in the atmosphere?

Because of those processes alone, the average carbon molecule stays in the atmosphere only about five years only about five years. That's a lot longer than the one week for a water molecule, but it's fairly short in the grand scheme of things. So why do we worry about excess carbon going into the atmosphere? Here's why—and this is the subtle point, the key point we need to understand—any carbon we put in the atmosphere joins that rapidly cycling system; cycling through the biosphere, the ocean surface waters, and back into the atmosphere. Of the carbon we put in the atmosphere, it turns out that roughly, if we put a kilogram of carbon or a ton of carbon in the atmosphere, about half of that ton ends up cycling through that system continually. So even though the CO_2 molecules we put in the atmosphere remain there only about five years on average, they are rapidly cycled back, at least about half of them. If we put a ton of carbon in the atmosphere, it turns out about half of it ends up in the soils and the ocean waters and stays there. We don't fully understand that breakdown, and that's an area of active research; it's called the carbon sink. Where does the carbon go?

The other half that we put in the atmosphere, even though it's pulled out by processes like photosynthesis, it quickly goes back. So, when we add carbon to the atmosphere, it stays there for a long time. The five years I talked about as being the typical residence time for carbon in the atmosphere is misleading. It's correct, but it's misleading because it suggests that, well, we put carbon in, it'll all be gone after five years. It won't be all gone because it will be participating in that cycle, and it will go right back into the atmosphere. The key idea is that carbon is going to stay in the atmosphere a lot longer than its five-year residence time. How much longer? That depends on two other processes that are involved in the very long-term removal of carbon from the atmosphere.

The two important processes both occur in the ocean, and they take carbon from the surface waters down into the very deep ocean, where it really has a hard time escaping. There are about 39,000 Gt of carbon in the deep ocean, a vast amount compared to anything that's anywhere else in the carbon cycle, and it has very little role to play, at least in this short-term climate cycle. How does stuff get down there? Well, one of the main drivers is something called the biological pump. If you're a tiny marine plankton, or even a fish or something, you eventually die, and you eventually sink to the bottom. You've got carbon that was ultimately fixed from atmospheric carbon by photosynthetic processes, whether you're a plant or an animal. When you sink to the bottom, you take that carbon with you, and that carbon is removed permanently from the rapidly cycling part of the carbon cycle. That's called the biological pump, and that takes down something on the order of about 10 or 11 Gt per year.

There are two other processes called upwelling and diffusion that bring up deep waters and bring some carbon back up into the surface waters. That brings up almost as much as the biological pump pumps downward—the biological pump being this decay of dead organisms—but not quite. There's a net removal of about two Gt of carbon to the deep ocean. At least on time scales of hundreds to thousands to hundreds of thousands of years, that carbon is lost. There is a net way of removing carbon from the system, but because that flow is so slow, it takes hundreds to thousands of years to remove carbon from the cycle. In fact, I've seen a recent study that suggests that some small fraction of the carbon that we might put in the atmosphere will remain for time scales up to 30,000 or 35,000 years. That's a long time.

If you say what is the residence time for carbon in the atmosphere, to be honest, I have to say it's five years, roughly. But then I have to add, but the carbon remains in the rapidly cycling system of the atmosphere, the biota, the surface waters, and so on for time scales—and this is where it gets ambiguous—but anything from hundreds to thousands of years is a good answer there. A nice even number is 1,000 years; some people use 300 years. There is no exact number for that. Some of it remains a lot longer; some of it shorter. But somewhere on the order of hundreds to a thousand or a few thousand years is the typical residence time for carbon; not in the atmosphere alone, but in that rapidly cycling system. That's the time it will take to remove any carbon we add from that system. That is the big, huge key idea here that you have to understand to realize why any anthropogenic carbon affects the carbon cycle, and the atmospheric carbon concentration, and therefore the concentration of greenhouse gases, and therefore the strength of the greenhouse effect for a long time.

There are two additional processes I should just point out that occur on much longer time scales. One is important in the natural climate cycle, and that is over long times, this carbon that sunk to the deep ocean formed sedimentary rocks. Those rocks are eventually brought up through tectonic processes. They're brought to the surface again where they can weather and release their carbon, or they're brought up through volcanoes, and the carbon is eventually released into the atmosphere through volcanism. Over the time scale of millions and hundreds of millions of years, geological processes complete the cycle much more slowly, and return that carbon that was moved to the deep ocean.

The one other part, which is not terribly significant naturally, but is terribly significant for human beings is that a very tiny fraction of the carbon that is removed from the atmosphere by photosynthetic plants, and maybe gets incorporated into animals or maybe stays in plants, decays and it dies and gets buried before it has a chance to decay and return its carbon to the atmosphere. So, it, too, doesn't participate in that rapidly cycling part of the carbon cycle, but instead remains buried underground, where it undergoes chemical processes, which eventually turn it into the fossil fuels; coal, oil, and natural gas. Those fuels have in them stored the solar energy that was fixed by photosynthetic organisms as they took that carbon out of the atmosphere and used solar energy to make it into new chemical compounds. That's where the fossil fuels came from.

We don't know exactly how much fossil fuel there is. One guess is there might be 10,000 Gt. There might be half that; there might be twice that. We don't know. If we knew, we would better be able to predict how we're going to do it in the next few decades. That's where the fossil fuels came from, and they play very little role in the natural climate system because basically, they just set there under the ground. What we human beings have done—and this'll be the subject of the next lecture—is a great deal with that carbon reservoir associated with the fossil fuels and that process.

Let me pause now. We're through the first half of the course, and I'm going to give you a very quick review of what we have learned in the first half of this course because now we're going to move to the anthropogenic part, the part about human beings and our role in climate. We know that the Earth is warming. We know that it's been warming especially much in recent decades. We've seen that from t he 150-year thermometer record. We've seen it from ice melt, from changes in weather patterns, from changes in species' behavior, from species' ranges, from springtime events, and so on. We know it from paleoclimate reconstructions, which suggest, for example, that the warming is unprecedented over the past millennium; that the longer term shows this periodic pattern of ice ages and warm spells; that a planet's climate is established by this balance of incoming solar energy and outgoing infrared radiation; and that balance is affected by the presence of atmospheric greenhouse gases that make it harder for the infrared to escape and therefore warm the surface.

For Earth, we know that water vapor and CO_2 cause a natural warming of about 33°C, and we know that there's a close correlation between temperature and atmospheric CO_2 concentration. Finally, we know that carbon cycles through the Earth atmosphere system; and as it cycles through the system, it does so in a way that, although the cycling is rapid and results in carbon remaining in the atmosphere itself only about five years, that cycling occurs rapidly, and the carbon is returned—or much of it—much of that carbon is returned to the atmosphere. It's only those very slow processes that bring carbon into the deep ocean that eventually remove it, and it takes those processes at least centuries, and in some cases thousands of years, to remove the carbon from the system. The bottom line of the first half of the course is CO_2 in the atmosphere causes a greenhouse effect, which warms our planet. We know that to be true and we also know enough about the carbon cycle to know that any carbon we put in the atmosphere is going to remain there for a long time.

Lecture Seven
The Human Factor

Scope:

For the past few centuries, we humans have been removing fossil carbon from the ground and burning it to provide energy. That process combines carbon with oxygen to produce carbon dioxide, which we release to the atmosphere through our tailpipes and smokestacks. About half that CO_2 accumulates in the atmosphere, with the result that atmospheric carbon dioxide has risen nearly 40% since the start of the industrial era—to levels the planet hasn't seen in at least a million years. We've released additional carbon through deforestation and agricultural practices, and land-use changes have altered the reflection of sunlight. All these processes affect Earth's climate, and recent climate change cannot be explained without including the human factor.

Outline

I. We begin with another quick review.

   A. The preceding lecture introduced the natural carbon cycle and showed how any additional carbon gets "stuck" in the rapidly cycling atmosphere–biosphere–surface ocean part of the cycle for hundreds to thousands of years.

   B. Earlier lectures showed how atmospheric carbon dioxide contributes to the natural greenhouse effect. Past climates show a strong correlation between atmospheric carbon dioxide concentration and global temperature. The science of the greenhouse effect shows why we should expect this correlation.

II. Now we explore the human influence on climate, concentrating first on our alteration of the carbon cycle.

   A. We're interested in this impact because human CO_2 emissions should lead to an *enhanced greenhouse effect* and, thus, to global warming.

   B. The dominant source of *anthropogenic* (human-produced) carbon in the atmosphere is the combustion of fossil fuels.

1. Fossil-fuel combustion results in about 7 gigatonnes (Gt, billions of tonnes) per year of carbon being emitted to the atmosphere in the form of CO_2. We can represent this anthropogenic CO_2 by an additional arrow in the carbon cycle, from the fossil-fuel reservoir to the atmosphere.
2. Additional emission of roughly 2 Gt of carbon per year results from deforestation.
3. About half the emitted carbon ends up adding to the atmospheric carbon dioxide concentration. The rest is taken up by biosphere, soils, and ocean waters.
4. Although 7 Gt/year is small compared with natural carbon-cycle flows (for example, 110 Gt/year for photosynthesis alone), the anthropogenic carbon accumulates in the system.
5. The result of anthropogenic emissions is an increase in atmospheric carbon from a pre-industrial value of about 560 Gt to about 800 Gt today.
6. The corresponding concentration increase is from a pre-industrial value of about 280 ppm to around 390 ppm. That's roughly a 40% rise. See Figure 10.

C. How well do we know the history of atmospheric CO_2?

1. CO_2 is easier to measure than temperature; air bubbles trapped in ice provide past concentrations.
2. For the past half century, we've had direct measurements of atmospheric CO_2. See Figure 10 detail.

D. How do we know that the CO_2 of the past few centuries is anthropogenic?

1. Fossil fuels are commercial commodities. We have a good estimate of how much has been burned, and the CO_2 increase is consistent with this.
2. Fossil carbon has been buried for hundreds of millions of years. Therefore, it's depleted in the radioactive isotope carbon-14. Furthermore, living things take up stable carbon-12 more readily than they do stable carbon-13, and living things are the origin of the fossil fuels. Therefore, fossil carbon is also depleted in C-13. Examination of the isotopic composition of atmospheric carbon shows that it is becoming increasingly depleted

in C-13 and C-14, indicating the fossil origin of the added carbon.

3. The oxygen content of the atmosphere is decreasing slightly, consistent with the oxidation of carbon through combustion.

E. How unusual is the industrial-era rise in atmospheric CO_2?

1. Going back 1000 years with ice-core data shows that the industrial-era rise is unusual in the millennial context and coincides with the recent unprecedented rise in temperature.

2. Going further back with ice-core data shows that the present-day CO_2 concentration is unprecedented over hundreds of thousands of years. Indirect evidence suggests that it is unprecedented in 20 million years. Quantitatively, today's CO_2 concentration is some 30% higher than anything the planet has seen in such long times. See Figure 11.

III. Anthropogenic CO_2 probably accounts for about half of the human impact on climate since the pre-industrial era. There are a number of other so-called *forcing agents* that also upset Earth's energy balance.

A. Methane (CH_4, natural gas) is another important greenhouse gas.

1. On a per-molecule basis, methane is about 26 times more effective in blocking infrared radiation than is carbon dioxide.

2. However, there's a lot less atmospheric methane, and its lifetime in the atmosphere is only about 12 years before chemical reactions remove it permanently.

3. Methane has many sources, both natural and anthropogenic. The latter include leaks from mining and natural gas transport, agriculture, landfills and sewage treatment plants, rice paddies, and even hydroelectric dams. The sizes of anthropogenic methane sources are less understood than those of carbon dioxide, but methane probably contributes about one-third as much as CO_2 to global climate change.

B. Other significant greenhouse gases include nitrogen oxides, halocarbons, and ozone (O_3). Ozone has a warming effect in the lower atmosphere but a cooling effect in the stratosphere.

C. Anthropogenic *aerosols*—airborne particulates—also exert an effect on climate.

1. Some aerosols result in warming, but the dominant aerosol effect is a cooling. The effect of most aerosols is more localized than that of greenhouse gases.

2. Especially important are *sulfate aerosols*, resulting primarily from coal combustion. These reflect sunlight and, therefore, decrease the inflow of solar energy, upsetting Earth's energy balance in favor of cooling.

3. Aerosols also influence cloud formation. Clouds are the least well understood aspect of the climate system, and this effect is highly uncertain. Some clouds absorb infrared, which results in warming, but the dominant effect of clouds is probably reflection of sunlight, resulting in cooling. This is the *indirect aerosol effect*.

D. Another important human influence on Earth's energy balance is land-use changes that result in changes in reflected sunlight. The net effect here is uncertain but probably entails a slight cooling.

E. For comparison, an important natural driver of climate change is variation in the Sun's energy output. This effect is small compared with the dominant human influences.

IV. Why do we attribute recent global warming to human activities?

A. Records of atmospheric carbon dioxide and temperature show a strong correlation, with both rising at unprecedented rates in recent decades.

B. More quantitative studies look at correlations between observed temperatures and the natural and anthropogenic influences on climate.

1. Natural influences include volcanoes, solar variability, and fluctuations inherent in the complex climate system.

2. Anthropogenic influences include greenhouse gases (warming) and aerosols (cooling).

3. Results show that natural causes can explain the pre-20th-century temperature record but that the dominant correlation in the 20th century is with anthropogenic causes.

C. Computer climate models are started from past conditions and run to the present. See Figure 12.

    1. Models that don't include anthropogenic effects satisfactorily reproduce observed climate into the mid-20th century.

    2. Only models that include anthropogenic effects can reproduce the climate of the late 20th century.

V. Let's close this lecture with a brief summary.

    A. Human activities, especially the burning of fossil fuels, are adding carbon dioxide to the atmosphere. This upsets Earth's energy balance in favor of warming. Today's atmospheric CO_2 concentration is nearly 40% above its pre-industrial value and is far above anything the planet has seen for nearly a million years and probably longer.

    B. Other anthropogenic greenhouse gases, aerosols, and land-use changes also contribute. Greenhouse gases produce a warming; the net effect of aerosols and land-use changes is cooling. The latter effects are quantitatively quite uncertain.

    C. Natural effects can explain climatic variations into the early 20th century, but only with anthropogenic greenhouse gases and aerosols can scientists explain the climate change of recent decades.

Suggested Reading:

Houghton, chapter 3, p. 42 to end.

Wolfson, chapter 14, section 14.3, "Attribution."

Going Deeper:

IPCC 4, chapter 9.

Harvey, chapters 7–8 (very technical).

Web Sites to Visit:

RealClimate, http://www.realclimate.org. This blog, maintained by scientists active in climate research, has technically accurate, detailed discussions of climate issues and, particularly, the attribution of climate change to human activities. Search the site on "attribution" for discussions relevant to this lecture.

Questions to Consider:

1. Human carbon dioxide emissions are a small fraction of the global carbon flows. Why, then, do we expect them to have a significant impact on climate?

2. In what way do pollutants emitted from coal combustion actually help to counter (but not prevent) anthropogenic global warming associated with greenhouse gases?

Lecture Seven—Transcript
The Human Factor

Welcome to the second half of the course; in particular to Lecture Seven, "The Human Factor." The title here is pretty much the theme for the rest of the course because we're now looking at what effect human beings have had on climate, and what our evidence is that we are in fact influencing the climate of today. Let me begin with a quick review of a couple points made in earlier lectures. In the previous lecture, you saw the carbon cycle—natural carbon cycle. You saw how carbon cycles between the atmosphere, living things, the oceans, the soils, and some of it is removed eventually into the deep ocean. You saw in particular a very important fact about the carbon cycle; namely that any additional carbon that gets into the ocean surface, atmosphere, biosphere, part of the carbon cycle, remains there for time scales up to hundreds to thousands of years. Even though carbon itself lives in the atmosphere for only about five years, the rapid cycling means that any additional carbon we put into that part of the cycle will be with us for a long time.

Earlier lectures showed also how atmospheric CO_2 levels and temperatures are very well correlated. What that leads us to believe is that, if we humans put additional carbon in the atmosphere, not only will it stay there for a significant time, but it will also cause Earth's temperature to rise. Without going into any more details at this point, I want to emphasize that that conclusion is basically physical science applied to a very simple system of the Earth and the atmosphere. Now, there are a lot of complications in that system. I can't tell you immediately how much the temperature is going to go up, but it makes perfectly good physical sense that, if we put additional carbon in the atmosphere, it will remain there for a long time, and it will cause Earth's temperature to rise. That rise in temperature associated with additional CO_2, particularly in the atmosphere—is called the enhanced greenhouse effect. Where does this additional carbon come from? The dominant source of the anthropogenic carbon—the carbon that comes from human activities—is from burning fossil fuels. That's not the only source of the carbon, and that's not the only source of the warming, as we'll see in the course of this lecture. But the dominant source is the combustion of fossil fuels.

Now I'd like you to think back to the diagram that I showed in the previous lecture of the carbon cycle. In that diagram, we had flows going between the atmosphere and the surface, between the surface and the atmosphere, between the plants and the atmosphere, and so on and so forth. We had one minor flow that I mentioned at the end, that was very insignificant, at least in terms of the quantity of carbon involved. That was the flow of carbon in the form of dead material that didn't have a chance to decay before it got buried, and eventually became the fossil fuels. There's a big reservoir of fossil fuels in the ground that contains a lot of carbon; about 10,000 Gt, we think. That reservoir, under the natural state of things, just sat there under the ground. We humans have begun tapping that reservoir as the source of most of our energy, as I'll show you in subsequent lectures.

What would you have to do to alter my picture of the natural carbon cycle? You'd have to add an arrow coming from that fossil fuel reservoir and going up into the atmosphere. How big is that arrow? That arrow represents a flow of about seven Gt of carbon per year. We human beings, through the burning of fossil fuels, are putting about that—or emitting—about seven Gt of carbon nto the atmosphere. We'd add another arrow to our diagram, an arrow representing the effects of deforestation. When we cut down forests, and the wood subsequently decays, and the absence of the forest means we aren't taking CO_2 out of the atmosphere as much, all that means roughly an additional two Gt of carbon is going into the atmosphere from deforestation. So that number is less certain than the number from fossil fuels, and I will give you reasons soon why we know so much about the fossil carbon. We're talking roughly an additional flow of nine Gt of carbon from the surface—or from the buried fossil fuel carbon in that case—into the atmosphere as a result of human activities.

That flow, seven Gt of fossil fuel carbon another two Gt from deforestation is pretty small compared to the magnitude of the carbon flows in that carbon cycle diagram; 110 Gt a year, for example, from photosynthesis. You might say, "That this is a minor effect. Why do we worry about it?" Again, that's why I went into such detail in the carbon cycle last lecture. The reason this is important is because any additional carbon we put in the atmosphere, any additions to those flows, remains in that cycling system for hundreds to thousands of years. So, it accumulates over time. If we keep putting seven Gt a year of carbon out, much of that accumulates in the atmosphere and continues to build up and build up. It isn't that we've perturbed the carbon flows in just a small way. While we have perturbed the carbon flows in a small way, but we've done it in a system where the additional carbon accumulates over time. We've been doing that since roughly 1750, roughly the start of the industrial era, when we started using fossil fuels wholesale.

What happens to that seven Gt of carbon that we put in from fossil fuel burning and the two Gt from deforestation? Roughly half of that ends up adding to the atmospheric carbon. The other half is taken up by a combination of the oceans and soils. Exactly where it goes and exactly how much is taken up is a matter of current research still. But of the carbon we put into the atmosphere, roughly half of it ends up basically staying there in the atmosphere. Again, it goes through that rapid cycling, but the atmospheric concentration goes up, the atmosphere concentration by about half the amount of carbon that we emit. As a result of that, we would have to make one other change in that carbon cycle diagram.

You may recall that the atmosphere was a reservoir of carbon with about 560 Gt of carbon before the start of the industrial era. Today— and I don't know when you're watching this course, so today may be one date or another, and this number is always changing—but roughly at the time you're watching or hearing this course—we have about 800 Gt of carbon in the atmosphere. That's a substantial increase from the amount there was before, in the pre-industrial era. That's again because even though the carbon flows associated with human activities are relatively small in the global carbon cycle, they put additional carbon into the system that continues to accumulate. The amount of carbon in the atmosphere has gone up substantially since pre-industrial times, and it's now about 800 Gt.

What effect does that have on CO_2 concentrations? That increases the concentration of CO_2 in the atmosphere as well, and the increase in CO_2 concentration shows a rise from about 280 ppm in the pre-industrial era, to something on the order of 390 ppm today—again, that depends on what today means for you, what day you're watching this. The amount of carbon in the atmosphere, or the concentration of CO_2 in the atmosphere, is rising today at about two ppm every year, give or take. Sometimes it's a little more; sometimes it's a little less, but it's about two ppm every year. At this point, it's somewhere on the order of 390 ppm. You'll remember my milk jugs; ppm simply means if we had a million gallon milk jugs of air, and we separated it out, in the pre-industrial era, 280 of those gallons would contain CO_2. Today, about 390 of them contain CO_2. That's a rise of roughly 40%. That's a significant impact we human beings have had on the CO_2 content of the atmosphere. Again, that couples with everything we know from the first six lectures, that we expect a rise in temperature because of the enhanced greenhouse effect, because of this additional carbon dioxide.

How do we know this history of atmospheric carbon? I want to tell you a little bit about that because we have considerable confidence in these measures of atmospheric CO_2. As I indicated earlier, it's pretty easy to measure atmospheric CO_2 in the past because we have bubbles of air trapped in ice cores, and that gives us a record way back. I showed you some records, and I'll show you them again, that go back some 500,000 years, roughly. In the modern era, we've done even better. Since about the 1950s, we've had monitoring stations at a number of places around the globe, and one of the most important ones is on Mauna Loa in Hawaii, sticking up 10,000 feet above sea level and sampling the clean Pacific air that isn't contaminated by being close to any major land mass. It's close enough to the equator that it samples air that's mixed from the Northern and Southern Hemispheres, and so, we believe, gives a good indication of the CO_2 concentration of the atmosphere globally. Let's take a look at our data on CO_2 concentration in the atmosphere from the beginning of the industrial era, from about 1750, to the present.

I have a graph here, which shows overall that rise in concentration from 1750 to about the present. Again, that's a rise from about 280 ppm to the present, about 390 ppm, roughly. Then there's a little inset that shows the data from the year about 1960, just after we started this monitoring program on Mauna Loa. This is data from the Mauna Loa measuring station. What I want to emphasize about that is that we measure this data so accurately that if you look closely at that graph, you'll see little up and down wiggles. Those aren't noise; those aren't random errors in the measurement. Those are the effects of the seasons. In the Northern Hemisphere, when spring comes, green plants come out—and the Northern Hemisphere has most of the land area of the world, so it has most of the green plants—those green plants, through photosynthesis, take CO_2 out of the atmosphere. As a result, in the spring and summer, the CO_2 content of the atmosphere goes down. Then winter comes, and the plants shut off much of their photosynthesis, and the CO_2 level goes back up again. That's what's reflected in those little fluctuations.

The reason I'm showing you this is to tell you that we can measure with exquisite precision the CO_2 content of the atmosphere, and it is indisputable that that CO_2 content has risen by something between 30 and 40% since the start of the industrial era. How do we know that that CO_2 is of anthropogenic origin? We have several ways of getting at that information. First of all, fossil fuels are commercial quantitites. They're things we buy and sell, so we know how much of them we're selling, very accurately. We know how much we're burning. We know how much we're putting in our cars. We have a pretty good sense of how much CO_2 got produced from the fossil fuels that we bought and burned. Consequently, we can keep track pretty accurately of how much carbon should have gone into the atmosphere from the fossil fuels that we burn in any given year.

Remember that only about half the carbon we emit from our tailpipes, and smokestacks, and so on ends up staying in the atmosphere. But we understand that also, even if we don't know exactly how that carbon is getting partitioned among the soils in the oceans that are taking them up. We have a good measure. If you tell me you burn a barrel of oil, I can tell you how much the CO_2 content of the atmosphere ought to go up. The measured industrial era rise in atmospheric CO_2 correlates with, in fact, our measured quantities of this commercial commodity, fossil fuel, that we've been burning. That's one way we know this well.

Another way is more scientific, and a little more subtle. Carbon, like plenty of other elements, comes in different isotopes, and there are three particularly interesting isotopes of carbon. Most of the carbon is carbon-12. It has six protons and six neutrons in its nucleus. That's the common kind of carbon. There's another isotope of carbon that's also stable—it constitutes much less of natural carbon—and that's carbon-13, which has an extra neutron in its nucleus. Then there's an isotope called carbon-14, which you may have heard of. It's a radioactive isotope. It only lasts about 5,000 years before it decays away, and it's created continuously in the atmosphere by cosmic rays. That's what's used in radio carbon dating, incidentally.

That means atmospheric CO_2 is a mix of carbon-12, carbon-13, and carbon-14—a very small amount of carbon-14, but it's easy to measure it because it's radioactive. The carbon in fossil fuels has been buried in the ground for millions, tens of millions, hundreds of millions of years. Any carbon-14 that was in it has long since decayed. So, it should be depleted in carbon-14. Furthermore, the plants that originally did the photosynthesis preferentially take up the lighter isotope, carbon-12, and so fossil fuels are also depleted in the stable isotope carbon-13. As we look at the isotopic content of the atmosphere—that is, what the ratio of carbon-12 is to the other two carbon isotopes—we find that over time, isotopically, the carbon content of the atmosphere has become more and more depleted in carbon-13 and carbon-14.

Why? Because we're putting into the atmosphere extra carbon, and the carbon is coming from a place where it had time to decay away its carbon-14, and where it didn't have as much carbon-13. So, the atmosphere carbon content is being reduced, depleted, in carbon-13 and carbon-14. That tells us that we're dealing here with carbon that was coming from deep within the Earth, particularly from the fossil fuels. Finally, one other measurement is telling us that the oxygen content of the atmosphere is decreasing in a way that's consistent with our combining extra oxygen with carbon, as we do when we burn the fossil fuels and make CO_2. You don't have to worry that you're about to suffocate. It's a subtle change, but it's a very clear measurable change, and it too is consistent with the addition of carbon to the atmosphere from the fossil fuel resource.

I think it's incontrovertible that the CO_2 content of the atmosphere has increased significantly during the industrial era, by something on the order of 30 to 40%. It's also uncontroversial that that rise is largely a result of our burning fossil fuels, which had been buried in the grounds for tens to hundreds of millions of years. We know we have anthropogenic carbon in the atmosphere, and a significant rise in atmospheric CO_2. But how unusual is the rise of approximately 30 to 40% over a period of 250 years or so, of the industrial era in terms of the context of historical CO_2? One thing I would remind you of is that we asked that same question for temperature. For temperature, we found that the temperature increase of the last few decades, at least, is unprecedented, at least on the millennial time scale, going back 1,000 years.

If you were to consider overlaying our CO_2 records for the past 1,000 years—and we can go back 1,000 years—what you would see is CO_2

was roughly level at that pre-industrial 280 ppm, approximately; and then about 1750, it started rising—and it took a very steep rise; the rise getting steeper and steeper as we get into the 20th century and into 21st century—that rise culminating in today's level of about 390 ppm. You would see, essentially, a level for most of the first part of the millennium, and then starting with the industrial era, a gradual upturn that gets steeper and steeper and steeper and the end. That's what a graph of CO_2 over the millennium would look like. It looks a lot like those temperature curves from those millennial temperature reconstructions that I showed you earlier. They showed a slight decline in temperature for roughly the first 900 years of the millennium, and then a steep upturn at the end.

If you were to imagine overlaying those temperature reconstructions on the CO_2 curve—millennial CO_2 curve, you would see quite a remarkable correlation in those two curves. The CO_2 rise during the industrial era corresponds with that steep rise in temperature at the end of the millennial temperature reconstruction. What if we go further back? I've already showed you the results of going further back with CO_2 measurements. I had correlated them with temperatures. I want to take another look at that, so I'm going to show you a graph, which is one that we've seen before, but I want to do one additional thing to it. I want to add the industrial era CO_2 rise to that graph. his, to me, is one of the most significant indications that we humans have done something substantial to the natural system of the atmosphere.

Let's take a look at what happens with 400,000 years of CO_2 from that Vostok ice core record. We remember that that correlated very nicely with the temperature, so when we put those two graphs on top of each other, they look almost the same. That's that correlation between temperature rise and CO_2 concentration. We saw that 6°C difference between the temperature today, roughly, and the temperature in an ice age. We saw that cyclic 100,000-year pattern with short interglacial periods of relative warm temperatures separated by long, cold ice ages when the climate was a lot cooler than it was today. By the way, the CO_2 level was down much lower, at somewhere around 140 or 150 ppm, much lower than it is today, a little below 200 ppm.

But if you add the present-day CO_2 concentration to that graph—and that's what I've done in the graph we're looking at here—add that up to 390 ppm, that addition looks like just a vertical spike at the end of the 400,000-year temperature and CO_2 record. A vertical spike, a rise in CO_2 to levels that the planet has clearly not seen in about 500,000 years; way above anything the planet has seen in 500,000 years; about 30% above any CO_2 level Earth has seen in 500,000 years. Our newest ice core data take that back about a million years, so we can say with confidence that we have added to the atmospheric CO_2, bringing it up to levels that are substantially higher than anything the planet has seen for a million years.

We have less direct evidence, but still good evidence, that probably the levels of CO_2 in the atmosphere today are higher by a substantial amount—something like 30%—than they've been anytime in the last 20 million years. To me, that's the biggest indication, again, that we human beings have altered the Earth atmosphere system—which is a major part of the climate system—in a way that is substantial and unprecedented. We're looking at a rise in atmospheric CO_2 to levels that the planet has not seen for at least a million years, and probably 20 million years, to levels that are well above anything its seen in those time intervals. That's a big deal. I think that's the clearest indication of a human influence on climate—or atmospheric composition, which ought to cause a human influence on climate.

CO_2 is not the only thing that we put in the atmosphere that changes climate. There are a number of other human activities that change climate, and I want to look briefly at them; some other human influences, basically on Earth's energy balance. Because anything that upsets that energy balance would change the climate, and I want to run quickly through these. As I describe these, think about what it would look like if one were to draw a graph of them. Some of these influences cause a positive upset in the Earth's energy balance. It makes it seem like more energy is coming in, or less energy is going out; it's the same thing either way. That's called a forcing, and a positive forcing like that corresponds to an increase in temperature, or would cause an increase in temperature.

Other things we do may alter the Earth's climate in a way that causes cooling, and we can describe those by imagining a bar that extends from zero downward, to some negative forcing value. Those forcing values are measured in watts/m^2. How much is the energy balance one every square meter, at the top of the atmosphere or at the stratosphere/troposphere boundary, how much is that energy balance

©2007 The Teaching Company

upset by these different phenomena? If we look at them, one of the more important is methane: methane gas, natural gas, CH_4 is its chemical formula. It's about 26 times more effective than CO_2 in producing climate change and producing warming. On the other hand, there's a lot less of it, and it stays in the atmosphere for only about 12 years.

It comes from many sources. We don't understand them as well as we do for CO_2. Leaks from natural gas systems, from transportation, from mining; agriculture produces lots of methane, rice paddies; cows belch methane; methane from a lot of sources—rice paddies, sewage treatment landfills, sewage treatment plants. Even some hydroelectric plants, particularly in the tropics, can actually produce more greenhouse warming than a coal-fired power planet because of the methane that comes from vegetation that's decaying on the bottom of the reservoirs. We don't understand fully again the balance of methane, but it probably contributes to about a third as much to global warming as does CO_2.

Other significant gases are nitrogen oxides, the halocarbons, the gases used in refrigerants and so on, and ozone, a pollutant which in the troposphere, actually causes the climate to warm; and in the stratosphere, actually causes a cooling.

But we do put other things into the atmosphere. We put aerosols into the atmosphere, a number of particulate matter; soot from burning coal, dust from industrial operations. But most importantly, sulfur from the burning of coal produces what are called sulfate aerosols, and they reflect sunlight and they exert a cooling effect on climate. Furthermore, they act as condensation nuclei for clouds and produce more clouds, along with the greater evaporation that's occurring, and that also reflects sunlight. Those processes are pretty poorly understood, even today, so we know that human factors also cause a cooling. We also know that that cooling is largely localized to the regions where these aerosols are produced. We also know that that cooling can't overcome the warming we're causing with our greenhouse gases. Nevertheless, there is a cooling effect there, and we have to take that into account when we consider what we're doing to the climate.

Just a few other things are land use changes that change the reflectivity of the planet. Although there is a balance of effects one way or the other, heating or cooling, the dominant effect, we believe, is a slight cooling. I'll just mention one natural phenomenon that also affects Earth's energy balance, and that's changes in the intensity of sunlight and the amount of power the Sun puts out. We have reason to believe the Sun's intensity has come up a little bit in the time we're talking about here, in the industrial era. But it's a much smaller effect than the human effects, and it is not a major contributor to global warming, although it does contribute slightly.

Why do we attribute the recent warming to human activities? I just described to you a whole host of human activities, from the burning of fossil fuels to changing land use patterns and so on, that should produce a net warming of the climate. Why, though, do we attribute the recent warming to human activities? First of all, there's that very strong correlation, which I've talked about and showed you, between CO_2 and global temperature. That continues today with that unprecedented temperature rise of the late 20th century and early 21st century, corresponding to a very rapid rise in CO_2. I should also tell you, we don't expect those two to go right hand-in hand, year-by-year, because there's a lag time before the CO_2 that gets into the atmosphere produces the warming. We don't expect a perfect correlation between those two. If you see some places where maybe it was cooling while the CO_2 was going up, that's not a contradiction of this idea. But overall, the global trend should be a correlation of those two.

More quantitative studies look at correlations in those reconstructions of the millennial temperature. For example, if you ask yourself: How much of this correlation is attributable to natural causes, particularly volcanic eruptions and variability in the Sun, and how much is attributable to human greenhouse gases? You can sort that out mathematically. What you find in looking at those temperature reconstructions is that, until the 20th century, the dominant explanation for what was going on, for the variations you saw, is natural factors, volcanism and solar variability. But then you find that, through the 20th century, the dominant factor in the explanation of those changes seems to be human-induced greenhouse gases.

I want to emphasize that discussion I've just given. That's all about observational data. Those millennial temperature reconstructions come from data—real data, from all that proxy data I described; coral

©2007 The Teaching Company

reefs, tree rings, ice cores, and so on. What I've just told you is that if you look just at data—this has nothing to do with climate models or any of that kind of thing—if you look just at data you find that you can correlate past temperatures with three things; volcanism, solar influences, and the human greenhouse gas emissions. Before the 20th century, the volcanism and the solar variability are sufficient to explain what you observe. But when you get into the 20th century, you need the human-induced greenhouse gases to explain what's going on. That argues strongly for a human influence on climate, at least in the 20th century and beyond.

We have also another completely independent argument from that one I just gave, for why we expect the human influence on climate, and why we expect that what's happening today is a result of human influence. We run computer models, and I'm going to say more about computer models in the next lecture, but computer models are large calculations that take into account the physics, the chemistry, the biology, all the atmospheric physics, the meteorology, everything that's going on in the climate system, and try to project what's going on. What I'm going to talk about are computer model runs that don't try to project what's going to happen in the future, but rather start about 100 years ago, move forward in time, and attempt to reproduce the present-day climate. The beauty of the computer models is there, we can do controlled experiments.

I'm going to show you a graph from some model runs; actually from an ensemble of model runs, so we're looking at an average of a number of climate model runs. As you look at this graph, the black curve, the dark curve, the wiggly dark line, the thin line, that is the temperature record—not exactly the same one I've been showing you before, but pretty much the same temperature records—from observations over the time period from just before the start of the 20th century to about the year 2000, in this particular case. There are two wider bands on this graph, and the wider bands represent an ensemble of computer model runs. They're wide bands because they represent an average. Each of these runs goes a little bit differently. All the runs were started with known conditions just before the start of the 20th century. The computer models were fed with those conditions, and they were asked to reproduce the climate through the 20th century.

They were asked to do so, in one case, with only natural factors included. We know about the volcanism; we know how much dust volcanoes have put in the atmosphere. We know pretty much how the Sun's energy output has evolved over this time. If we put those in, we get the lower of the two bands that you see on this graph. They're not terribly distinct through about the early part of the 20th century, or even the first half of the 20th century. What that tells us is that either of those two sets of model runs does a pretty good job of explaining the climate, at least through the first half of the 20th century. The upper band represents climate model runs that include those natural factors, volcanism and solar influences, but also includes the anthropogenic greenhouse gases, and also the cooling effect of those aerosols I talked about. We think we've got most of the human influence included in these model runs.

It's very clear, after the second half of the 20th century in these model runs, that only the upper band—only the band that includes the anthropogenic and the natural effects—reproduces the observed temperature. This is a very clear indication that we can't explain that temperature rise, at least not the rise of the late 20th century, unless we take into account the presence of anthropogenic effects, particularly the presence of greenhouse gases that tend to warm the climate; but also the effect of the anthropogenic aerosols that tend to cool the climate. That's all in here, and I think you'll agree, looking at this graph, that the model runs that include the human influence do a pretty good job of reproducing the entire temperature record, including that steep unprecedented rise of the last part of the 20th century.

Let me give a quick summary of this very important lecture. Human activities, especially burning fossil fuels, are adding CO_2 to the atmosphere, and Earth's energy balance is then upset in favor of a warming. In fact, today's CO_2 concentration is some 40% above what it was before the start of the industrial era. There are other human contributions. There are other greenhouse gases, there are aerosols, there are land use changes, there are contrails from jet aircraft—I didn't even mention them. There are other factors that are also influencing climate, and we believe we've taken them into account. The greenhouse gases in particular produce a warming; some of the other factors produce warming, some produce cooling. But we're taking them into account, and I will emphasize they're quite a bit less certain than the greenhouse influences.

©2007 The Teaching Company

Finally, the natural effects can explain climate variation to at least the 20th century, and maybe well into the 20th century. But if you want to explain the climate of the last few decades, you can't do it unless you take into account these anthropogenic effects. Human beings have had a real influence—and are having a real influence—on Earth's climate.

Lecture Eight
Computing the Future

Scope:

Climate models explore how climate behaves in response to human-induced changes and natural factors. Some models are simple schemes that look only at global average temperature. Others are complex, dividing Earth's surface, atmosphere, and oceans into small cells and computing physical, chemical, and biological processes in each cell, as well as exchanges of matter and energy among cells. Models also incorporate *feedbacks*, whereby a change in one factor leads to effects that can either exacerbate or mitigate the original change. Climate models make a range of projections, but most suggest that we can expect a global temperature rise of several degrees Celsius over the next century.

Scientists have confidence in their models because they successfully reproduce today's climate starting from past conditions; because they predict climate change resulting from volcanic eruptions; and because they reproduce patterns of precipitation, barometric pressure, winds, ice and snow, and other observed features of Earth's climate. But we still don't understand every detail of climate, and we can't know future human behavior; thus, models cannot give us precise predictions of future climate. Nevertheless, different models agree that we should expect a general warming, and they agree on many details, as well.

Outline

I. *Climate models* are mathematical descriptions of Earth's climate system. They provide projections—not predictions—of future climate given a set of starting conditions and assumptions about future emissions of greenhouse gases and other factors. There's a hierarchy of models, increasing in complexity. All are useful for understanding the climate system.

   A. The simplest models treat Earth as a single point and provide only a global average temperature, as in the simple calculation of Lecture Four.

   B. Two-box models treat surface and atmosphere separately, accounting for the flows of energy between the two, as well as energy input from the Sun and infrared energy radiated to space.

C. More complex models include variations with latitude and may have many layers of ocean depth and height in the atmosphere.

D. The most complex models, called *GCMs* (for *global climate model* or *global circulation model*), include realistic descriptions of Earth's continents and oceans in three dimensions. They divide Earth's surface, oceans, and atmosphere into many small *cells* and account for flows of energy and matter among the cells.

   1. There's a fourth dimension: time. Some models provide only *equilibrium* climate projections, while others track the evolution of climate change through time.

   2. *Coupled models* use separate modules for different aspects of the climate system—oceans, the carbon cycle, the cryosphere, and so on. They're coupled through the exchanges of matter and energy between the different modules.

II. Models aren't perfect, for a variety of reasons.

A. Their *resolution* is limited.

   1. Even the best global models have grid cells about 100 miles on a side, and they may divide oceans and atmosphere into 20–40 layers. That makes for several million cells, each with many equations describing processes in that cell and its interactions with adjacent cells. Those equations must be solved repeatedly as the model steps forward in time.

   2. Global climate models tax even the fastest computers, which can advance the model climate some 5–10 years in 24 hours of computation time. That's one reason simpler models are often used for exploring ranges of future climate scenarios.

   3. *Moore's law*—the exponential increase in computing power that has computer speed doubling roughly every 18 months—allows ever finer resolution and gives us ever more powerful and accurate climate models.

B. We don't understand the details of every climate process.

For this reason, climate models aren't based entirely on first principles but include some processes that are modeled on the basis of observational data.

C. Some climate processes occur at scales below the grid-cell size. These are handled by approximate schemes called *sub-grid parameterization*.

   1. Clouds such as cumulus are much smaller than the grid size. A crude way to handle a partly cloudy day is to assume a uniform cover of partially transparent clouds over a whole grid cell. More sophisticated parameterizations account for the vertical structure of clouds and the presence of clear sky between clouds, without actually resolving these features.

   2. Such features as cities, lakes, forests, and agricultural lands may be smaller than the grid size. Their different climatic effects must also be parameterized.

III. *Feedback effects* are an important aspect of the climate system, and climate models must incorporate these effects.

A. A feedback effect is an additional change that occurs following some change in a system.

   1. If the additional change enhances the original change, then the feedback is a *positive feedback*.

   2. If the additional change opposes the original change, then the feedback is a *negative feedback*.

B. Many feedbacks operate in the climate system. Three of the most important include:

   1. Ice-albedo feedback, in which a warming results in the melting of ice and snow, both of which reflect sunlight. (*Albedo* means reflectivity.) The darker land or ocean that's exposed absorbs more solar energy, resulting in still further heating. Ice-albedo feedback is one of the reasons why polar regions show the greatest temperature increases. This is a positive feedback.

   2. Cloud-albedo feedback, in which warming results in increased evaporation and, thus, more clouds. But clouds reflect incoming sunlight, lowering the energy input to Earth and countering the initial warming. This is a negative feedback.

3. Water-vapor feedback, in which warming results in increased evaporation and, thus, more atmospheric water vapor. Because water vapor is a greenhouse gas, the effect is additional warming. This is a positive feedback and is estimated to increase the effects of anthropogenic greenhouse warming by some 50%.

C. Some feedbacks may involve human behavior. For example, in response to warming, people buy more air conditioners. That means more coal is burned to generate electricity, and that means more carbon dioxide emission. That further enhances the greenhouse effect, leading to more warming. Thus, this is a positive feedback.

D. Although negative feedbacks counter a warming effect, it's possible to show mathematically that they can reduce the effect but not reverse it—at least for situations in which the climate responds gradually (linearly) to changes.

E. Feedbacks work in either direction. For example, a cooling of the planet results in more ice and snow, thus more reflection of sunlight, and thus, further cooling. This is, again, the positive ice-albedo feedback.

IV. *Model validation* procedures give scientists confidence in their climate models.

A. Climate models started with past conditions successfully reproduce present-day climate, as discussed in the preceding lecture.

B. Climate models reproduce many details of Earth's climate, including:

1. Temperature structure of the atmosphere.
2. Geographical distribution of precipitation.

C. Future climate projections show consistency.

1. All show temperature increases with increasing atmospheric CO_2.
2. However, projections show a considerable range in values of projected temperature changes.

D. We can't experiment with Earth's actual climate, but nature can—in the form of volcanic eruptions that inject enough dust into the atmosphere to affect global climate. Climate models successfully describe the resulting impact on global temperature. See Figure 14.

Suggested Reading:

Houghton, chapter 6 through p. 128.

Wolfson, chapter 15, sections 1–2.

Going Deeper:

Ruddiman, chapter 3.

IPCC 4, chapter 8.

Web Sites to Visit:

Climate Prediction Experiment, http://www.climateprediction.net. Participate in a worldwide climate modeling experiment! ClimatePrediction.net harnesses the spare computing power of personal computers worldwide to run multiple simulation experiments using the world's most sophisticated climate models. You can sign your own computer up to participate and keep yourself posted on the results of the ClimatePrediction experiments.

Questions to Consider:

1. Why is it useful to have a hierarchy of climate models, from the simplest to the most complex computer models?

2. How did the Mount Pinatubo eruption give climate modelers a natural experiment with which to test their models?

Lecture Eight—Transcript
Computing the Future

Lecture Eight: "Computing the Future." I've already argued we should expect any additions of CO_2 to the atmosphere to result in global warming. The temperature should go up because of the blocking of outgoing infrared radiation by an enhanced greenhouse effect that goes above and beyond the natural greenhouse warming that's already there of 33°C. But how much should we expect the temperature to go up? That's a more complex question because the climate system is a complicated system. Our tools for answering the question, what would future climate look like, are mathematical climate models that describe as much as we can about the processes that operate in determining Earth's climate. I want to describe those models in this lecture—to give you a sense of what they do, how they do it, and in particular, why we believe they give us an accurate or semi-accurate picture of what the future brings.

I'm not going to say that these models predict the future. These are not like weather predictions. The best we can do in weather predictions, by the way, is about 10 days to two weeks out. Beyond that, things get chaotic enough that we simply can't predict the weather in advance. Nobody's attempting to predict the exact climate. What we're trying to do is project the general climatic conditions that we can expect in the future under a variety of scenarios for what we human beings might do in terms of, for example, greenhouse gas emissions, changes in land use, changes in air pollution regulations that might cut down on the amount of aerosols and actually cause more warming because we reduce that reflection from aerosols. All kinds of possibilities we need to consider. We need to be able to put those into models that allow us to project what might happen under different conditions. Our models become like our laboratories. For climate scientists' climate models are the laboratories in which they do experiments because, as I argued earlier, we can't do real experiments—controlled experiments—with Earth's climate.

Let me tell you a little bit about some of the climate models we use. First of all, there are a range of climate models, from very simple ones that give you quick answers, that may be somewhat accurate in describing a limited amount of what the future climate would look like, to very complex models that incorporate virtually every process we understand. You might think, "Well, in this day and age of very fast computers, all we really need to do are these complicated models, but actually that's not true. It turns out that the whole hierarchy of climate models, from the simplest to the most complicated, is useful in various ways. The simpler ones can elucidate very obvious physics, chemistry, or ideas about what is happening in a particular situation; whereas a complex model gets so complex that it's like the actual Earth system, and it's a little bit harder to interpret sometimes what it has to tell us. The simpler models can run very rapidly, and therefore you can do many, many experiments with them; whereas, as I'll soon show you, the big complicated models take a very long time to run, even on the fastest computers.

I want to begin describing this hierarchy of models and kind of building them up in complexity. We've actually used climate models already in this course. Early on, I described how you could calculate Earth's temperature in the absence of an atmosphere by simply equating the 240 watts/m^2 of sunlight coming in to the known expression involving fourth power of temperature (T^4), that tells us the rate at which something that's hot, at a given temperature T, radiates electromagnetic energy. That equation alone—equating that outgoing infrared energy as a function of temperature to the incoming 240 watts/m^2 of sunlight—that was enough to give us a global average temperature.

That was a very simple model; it was a zero-dimensional model. I treated the Earth as just a single point. It didn't distinguish land from water, surface from atmosphere. In fact, it didn't even have an atmosphere. It didn't talk about poles versus equator. It treated the Earth as a single point, but it came up with a not unreasonable answer for the Earth's average temperature. When we understood the greenhouse effect and took it into account, we could add that to that simple model. We'd sort of be accounting for an atmosphere that way, and we could get a fairly accurate estimate of the Earth's global average temperature, and we could experiment with what would happen if we enhanced the greenhouse effect and saw what the temperature did. Even that simple zero-dimensional model tells us something.

Back in Lecture Four, I showed you a diagram of all the complicated energy flows between Earth and atmosphere that comprise Earth's

©2007 The Teaching Company

climate system. That complicated diagram still didn't take into account variations in latitude, and longitude, and other aspects of the Earth's surface. It basically treated everywhere on Earth's surface as being the same. But it did take into account the interactions between the Earth and the atmosphere, interactions that are very important in establishing the greenhouse effect, at causing reflection, and so on. A still very simple model for climate is a so-called two-box model, in which you treat the surface as one system, the atmosphere as another system, and you take those various arrows you saw in that diagram—that was a diagram based on observations; it wasn't a model—you take those arrows and the diagram and you quantify them with equations that describe how much energy is emitted by the surface at a given temperature, how much energy is emitted downward from the atmosphere when it's at a given temperature, and so on.

You would end up sort of redrawing that diagram with two boxes, and showing arrows representing, for example, sunlight coming in from outer space, reflected sunlight going back to space, infrared emission from the atmosphere, infrared emission from the surface to the atmosphere, infrared emission from the atmosphere to the surface, convection and evaporation going up from the surface and carrying energy into the atmosphere, sunlight that comes through the atmosphere and carries energy down. You could put arrows in representing all those features that appeared in that diagram of the climate system. You could then solve, simultaneously, the equations that describe all those flows in terms of things you know, like the amount of incoming sunlight, 240 watts/m^2; the amount of greenhouse gases, which tells you how much infrared gets absorbed; and so on. You would come up, in that two-box model, with a temperature for the atmosphere and a temperature for the surface.

When I teach my global climate change course at Middlebury College, a kind of intermediate undergraduate-level course—it requires some mathematics—my students actually build a two-box climate model, and they put the equations into a computer or, because this model is simple enough, they actually solve them with pencil and paper. They get out an atmospheric temperature and a surface temperature. In this model, where the entire atmosphere is lumped together, one thing they immediately see is the atmosphere has to be cooler than the surface. They can experiment with changes in the amount of greenhouse gases and changing the numerical quantities represented by that in the model, and they can experiment and look at how the surface temperature and the atmospheric temperature change, for example. That's a two-box model, and that's a fairly sophisticated, but nevertheless still simple, way of looking at the climate system and making some serious projections about what might happen under certain circumstances.

When the models get more complex, they tend to include variations with latitude, and they may have many layers of ocean depth, or many layers in the atmosphere. Those are called two-dimensional models, models in which things vary in two different directions. In a two-dimensional model, for example, typically you vary in altitude, and you vary in latitude because latitude is far more significant in determining conditions on the surface of the Earth than longitude; because the Earth is basically symmetric around its rotation axis, but it's very unsymmetrical from poles to equator and back to poles again. A two-dimensional climate model is more complicated still; it involves typically variations with latitude and variations with altitude. You can show in such a model things like the large-scale circulation of the atmosphere, which results from the heating of the surface, typically, mostly at the equator. Air rises, and it forms large cyclic patterns of air movement that, among other things, transport moisture and energy poleward, both toward the North Pole and toward the South Pole.

A two-dimensional climate model can do quite a bit for you in terms of exploring the processes that go on in the climate system, but the more complex models are three-dimensional. They're called GCMs, which you might think stands for global climate model—and today it often does—but it's an outgrowth of the acronym for global circulation models because these models looked at the entire globe in three dimensions and attempted to calculate climatic effects, including the circulation of the atmosphere. How do these models work, these GCMs, these global climate models or general circulation models? What they do is divide Earth's surface, and atmosphere, and ocean into small regions called cells. Within each of those cells, they calculate the processes that are going on that establish climate.

What are those processes? They're typically flows of energy from adjacent cells. Maybe if you have a cell at the bottom layer of the atmosphere, energy is coming up from the surface, energy is also

©2007 The Teaching Company

coming down from the Sun; there's infrared energy. You also calculate the movement of material in and out of these cells. Maybe there's an airflow occurring, and it's bringing heated air into the region, or maybe there's moist air coming into this region. You calculate the flows of energy and matter, and also momentum, which gives you forces on adjacent cells. You calculate the flows of those fundamental physical quantities into and out of these cells. You do that for all the cells that represent the layers of the atmosphere, the layers of the ocean, the different parts of the Earth's surface, and in very realistic models, you include continents and oceans, topography like mountains, and you do an enormous calculation, solving many, many equations simultaneously. Out of that comes a model climate. Of course, you can alter the model's atmosphere to see what would happen if we put more CO^2 in the atmosphere; what would happen *if* we stopped greenhouse emissions today, and so on. These models represent serious attempts to look at the structure of the entire climate system by breaking the Earth, the atmosphere, and the ocean up into many, many tiny cells and calculating what's going on in each of those cells.

Now there's a fourth dimension; and the fourth dimension is time. Some of the models, even three-dimensional ones, are not time-dependent. In other words, they only calculate what the equilibrium climate would be. You make some changes; you recalculate the model, and that tells you, if you waited a long time, this is the climate you would end up with. As I mentioned earlier, there are lag times; for example, lag times between when we put carbon in the atmosphere and when we see the effect of that on climate. The most sophisticated models are also time-dependent models, and they calculate as time goes on. They advance in time, doing a big calculation of this entire model—all these cells representing the globe, and the atmosphere, and the ocean—time and time and time again. Therefore, they give you a picture of how climate changes over time. Those are the most sophisticated of the models.

Many of these models, by the way, use specially developed modules for different parts of the model. For example, the oceans may be treated differently in a model that oceanographers have developed from the atmosphere in a model. What climate modelers do with these large-scale models is take an ocean module, for example, which is known to calculate ocean properties well, and they couple it mathematically, computationally, in their computer model at the surface of the ocean where it joins the atmosphere. The ocean model couples to the atmospheric model. Those are called coupled GCMs—AOGCMs, atmosphere/ocean coupled global climate models, or atmosphere/ocean coupled general circulation models.

These models are not perfect. They're not perfect for a variety of reasons. They're not bad, but they're not perfect. One of the most important reasons is that their resolution is limited. By resolution, I mean how finely these models divide the Earth, the ocean, and the atmosphere into small regions. Even the best of the global models today have grid cells that are only maybe 100 miles on a side. Maybe a piece of land area 100 miles by 100 miles is what they're representing as one unit. They don't get down to any details smaller than that. They divide the oceans and atmospheres into maybe 20 to 40 layers. That's pretty good, but it's not perfect because these are really continuously varying regions.

Even if you do what we do today, grid cells 100 miles on a side and 20 to 40 layers in the atmosphere and ocean, you've got several million of these grid cells in your model. Each of these grid cells probably has dozens and dozens of equations that have to be calculated to describe the flows of matter, and energy, and momentum into and out of that grid cell. You can imagine that you have an enormous computational task. These models—except for the simplest ones like the zero-dimensional model I described earlier, or maybe that two-box model that can be calculated by pencil and paper—all the rest of them have to be done on computers. The GCMs we have today, the biggest ones, tax even the fastest computers we have. They tax them to the extent that, in 24 hours of actual computing run time, a day of actual run time—a full day of computer running 24 hours—they can advance the climate maybe five to 10 years. So if you want to explore what the climate is going to do for 100 years, you may have to run this model for 20 days, for the better part of a month. If you want to really ask long-term questions about what's going to happen, you may have to run for months and months and months. There are climate model runs that have taken the better part of a year, and this is a big problem.

Climate computer models tax even the fastest computers. There's hope for the future, and that hope lies in Moore's Law. If you have seen my Teaching Company course *Physics in Your Life*, when I talk about semiconductor engineering and the development of computers and so on, I talk about Moore's Law, which is sort of an empirical statement that says computing power doubles about every 18 months. That's rapid exponential growth, and that's why the personal computer you bought yesterday is already obsolete, because computers are advancing that fast. There's hope for the future. Our climate models are improving in resolution, and the ability to calculate for long times into the future, because of these advances in computer engineering that are bringing us ever-faster computers. That means, with faster and faster computers, we can make the resolution finer and finer, and the models become generally more accurate that way.

There are other reasons the models aren't perfect. We don't understand the details of every climate process, for example. Some processes that are in there, like for example the radiation of infrared energy from a hot surface, we know precisely. We understand the physics of that perfectly; it's based in fundamental physics. But other processes, particularly processes involving clouds—the transfer of energy through clouds depends on the size of the droplets in the cloud, the condensation nuclei on which those droplets condense, the structure of the cloud. The reflectivity depends on all those things. We really don't understand clouds in all that detail. Clouds and some other processes also have to be sort of put in empirically. We look at what the real world does; we make a kind of empirical equation that reflects that, and we put that in our climate models. We don't understand all the processes from first principles. It would be great if we could just throw a few laws of physics into a model and say compute the climate, but we're nowhere near there yet. That doesn't mean the climate models aren't valid; it simply means they aren't all based entirely on first principles.

Another problem is that some of the processes we want to describe happen at size scales below that of these grid cells. For example, if you have a city, the city is probably substantially less than 100 miles across. If you have a 100 x 100-mile grid cell, and you want to include the climatic influence of that city, which has the effect of raising the temperatures—because of this urban heat effect I talked about earlier—how do you include that? You somehow have to smear that city out over that whole 100-mile grid cell. For example, a particularly important phenomenon that occurs often at scales smaller than the grid cell size is clouds. Imagine a sort of semi-fair weather day when you've got a bunch of little puffy cumulus clouds. Those are a lot smaller than 100 x 100-mile grid size.

Suppose the atmosphere is half covered with these small cumulus clouds. You might say okay, I can model that in my model by saying the entire grid cell is covered with clouds, but they're only 50% opaque. That sort of achieves the same thing. The same amount of solar energy gets to the ground as it does if half the sky is covered by completely opaque cumulus clouds. But if you think about that a little bit, that's physically rather different. A thin overcast that blocks out half the sunlight is a very different condition than having regions of clear sky with direct sunlight coming through, separated by regions where clouds are completely blocking the sunlight. In fact, you can think of some phenomena that might occur. Sunlight can actually bounce off the vertical sides of those clouds and be scattered downward. You can actually get regions where the intensity of sunlight is a little higher than it would normally be in clear air because of that effect.

Parameterizing the presence of those clouds—it's called parameterizing, sub-grid parameterizing—by assuming a uniform overcast with the same total amount of sunlight coming through just doesn't cut it completely. It's a first approximation, but we've got to do better. So, a lot of these climate models include sophisticated so-called sub-grid parameterizations that take into account processes that are occurring on spatial scales that are smaller than the grid size. Forests, lakes, these agricultural regions, or other areas that might have to be handled with sub-grid parameterization.

One of the most important aspects of the climate system, and one that's incorporated into the models—we think correctly, in most cases—are feedback effects. What is a feedback? Here I am teaching a class, and if I ask you to give me some feedback—unfortunately, it's not an interactive class, so we can't—I say "Hey, give me some feedback on my teaching." You say "Hey, Wolfson, you're talking too fast." Then I might slow down a little bit and talk a little slower. That's feedback; it's a corrective effect. I ask for some correction, and you give me the correction, and my behavior changes as a result. That's a feedback that is corrective. Some feedbacks are corrective; other feedbacks exacerbate the original change. Feedback effects are additional effects that either enhance or diminish the effects of some original change. That's a sort of cumbersome way to describe it.

If the additional change enhances the original one, that's called a positive feedback. In the climate system, if something makes the climate get warmer, and that causes some additional effect that makes it get warmer still, that's a positive feedback. When I talked about the correlation of CO_2 and temperature over the paleoclimate record, over those 400,000-year ice core records, I said there were feedback effects, positive feedback effects. A little more CO_2 came into the atmosphere and made it a little warmer. That caused other effects that made more CO_2 come in and made it a little warmer. We had a leapfrogging there of positive feedback. On the other hand, if the additional change opposes what was going on at first, then that's called a negative feedback; that's the self-correcting. If I start to talk faster and you scream at me, hey, you're talking way too fast, that's a negative feedback. It corrects, it diminishes, the original effect. But, as I'll show you in a moment, it doesn't necessarily correct it completely.

There are a lot of feedbacks operating in the climate system; many, many feedback effects. They involve all kinds of factors, from biological organisms to physics, to chemistry, to geology. All kinds of feedback effects operate in the climate system. One of the most important is called ice-albedo feedback. Albedo is a word that means reflectivity. The Earth's albedo, for example, is about 30%, about 0.3, which means about 30% of the sunlight coming to Earth is reflected back into space. Different parts of the Earth have different albedos. The oceans are dark; they absorb sunlight. They don't reflect much, and they have a relatively low albedo. Ice sheets and snow are very reflective; they have a very high albedo. The land surface depends on what it's made of and what the vegetation looks like. Forests have fairly low albedos; they absorb a lot of sunlight, and they aren't very reflective. Deserts have a fairly high albedo; they reflect a lot of the incident sunlight.

Let's talk about ice-albedo effect. The ice-albedo effect is a positive feedback in which, particularly in regions where there's ice and snow, if it warms up a little bit, then some of that ice or snow melts. If enough of it melts to expose either dark ground or, more significantly, dark open water, then that diminishing of the ice cover causes more dark surface to show; the albedo goes down. That causes more sunlight to be absorbed. That causes more heating, and that exacerbates the effect because it causes more ice or snow to melt. We get this cyclic positive feedback, the ice-albedo feedback.

The ice-albedo feedback is a really important feedback mechanism in the climate system, and it's the reason, as I've stated several times, why we expect, and are seeing, that the Arctic regions warm a lot faster than the rest of the globe, and why we're seeing particularly dramatic climate change in the Arctic; particularly the Arctic, and not so much the Antarctic, because the Arctic ice, most of it, is floating sea ice. It overlies dark water, and as we reduce the coverage of sea ice, we increase the absorption of solar energy because we've reduced the reflectivity, and that further exacerbates the heating effect. Ice-albedo effect is a very important positive feedback effect.

There's another effect; there's a cloud-albedo effect. The cloud-albedo effect is opposite; it's a negative feedback. What happens? Suppose there's a little bit of an increase in Earth's temperature. One thing that does is cause more evaporation. If there's more evaporation, more clouds form. Clouds, at their tops, have a high reflectivity. They reflect sunlight, and they therefore reduce the amount of sunlight getting into the climate system, and that causes Earth's temperature to go down. That's a negative feedback. You might say, "Well, we can stop global warming with these negative feedbacks." But if you look at feedback effects, especially those that occur sort of in proportion to the effect that's originally driving them, you can show mathematically that the best a negative feedback effect can do is to reduce the initial warming, but it can never turn it into a cooling. It can never reverse the initial thing that started out. The most it could do, if it were an infinitely strong feedback effect, would be to halt that warming all together. But no feedback is like that, so the best you can do with negative feedback is simply to reduce the initial effect, but you can't eliminate it or reverse it. That's cloud-albedo feedback.

There's another important feedback called water-vapor feedback, also related to evaporation. Water-vapor feedback is a positive feedback. If the Earth warms, there is more water vapor in the atmosphere, but water vapor is a greenhouse gas, and so there's a bigger greenhouse effect. Water vapor is a significant greenhouse gas. We believe today that global warming caused by CO_2 emissions is enhanced by about 50% by the water-vapor feedback effect.

Because water vapor goes in and out of the atmosphere so instantaneously, we don't regard water vapor as a greenhouse gas we put there; we regard it as something that instantaneously adjusts to the current climate conditions. But there is this water vapor feedback, and that probably enhances any warming from other factors by about 50%.

Some feedbacks may even involve human behavior. Suppose it gets real warm, for example: lots more people go out and buy air conditioners. They plug them into the wall. If they live in a region where their electricity comes from fossil fuels, they then produce more CO_2, which causes more warming. That's probably a trivial one in the grand scheme of things, but it illustrates that we human beings can have a role in these feedback effects, and therefore altering the climate system.

I want to emphasize that feedbacks work in either direction. The ice-albedo feedback—if the Earth began to cool, that would cause more ice to form. That would reflect more Sun, and that would cause it to cool further. That's still a positive feedback. It doesn't matter whether the initial effect was in one direction or the other. If the effect is enhanced, it's still a positive feedback.

I've talked a lot about computer models, some of what they contain, some of their imperfections, some of the things we're doing to make them better. How do we know that these computer models are any good? We have considerable confidence, actually, in the models. Climate scientists are doing a lot of things to test their models. First of all—and I showed you one example of this in the previous lecture—if you start a climate model in the distant past, 100 years ago or a couple hundred years ago at the start of the industrial era— some of them started much further in the past—and you run them to the present, you can make them reproduce present conditions, at least if you put in everything we know about what's driving the climate system. Climate models started in the past can successfully reproduce the present-day climate.

They can also reproduce some other pretty obvious things. For example, if you had a climate model that included the tilt of the Earth's axis, and the seasons and so on, it better be able to reproduce seasonal temperature variations, or you want to throw that model out. Models do reproduce things like seasonal temperature variations easily. Climate models reproduce also a number of other details of Earth's climate system that you might not think about. For example, I mentioned that climate models have many boxes representing the different layers of the atmosphere. We aren't just concerned about surface temperature. If a climate model reproduces the climate accurately, it ought to reproduce the temperature at all levels in the atmosphere. If you run a whole bunch of different climate models and compare their results for the vertical structure of the atmosphere, averaged over the planet, to what we actually observe for that vertical structure, you find remarkable agreement between the climate models and the observed vertical structure of the atmosphere. There's a subtle thing that you might not think about that the climate models very successfully reproduce.

If you look at patterns of precipitation, for example, how does precipitation depend on latitude? You average over longitude because longitude is a less significant variable in terms of climate; there again, latitude equator to pole. Precipitation is a more difficult and subtle thing to deal with. You find, nevertheless, that the computer models do a very good job of reproducing the latitudinal distribution of precipitation; they all show a big rise in precipitation at the equatorial regions. There's a drop in the subequatorial regions, like where the Sahara Desert is, and then a rise into the temperate zones, and then a fall-off as you go toward the poles. There's another example of something rather subtle you might not have thought of that those climate models—many, many different climate models by many different groups—are nevertheless successfully reproducing.

That's about current climate and how the models reproduce various aspects of current climate. If you project into the future, one of the reassuring things is that the models from different climate modeling groups working independently, having written different models with different sub-grid parameterizations and different other features do a pretty good job of consistently reproducing the future climate, or agreeing on what the future climate should be. A common experiment that climate modelers do is to subject their model to 1%/year increase in the amount of CO_2. If you look at a dozen or so different climate models—again, from around the world—and how they do at that rise in temperature with increasing CO_2, most of them cluster very close to the same average value, which comes out somewhere between 1.5 and 2°C rise at the end of a period of about 80 years. A couple of the models are a little further outlying. They all show an increase; none of them show a decrease. The increase they show is pretty much consistent.

Models are verified in a great many ways. What we'd really like to do with our models is say okay, let's do an experiment with the real climate system and see if our model can reproduce the outcome of that experiment. That, as I've argued many times, we can't do. However, Earth does provide us with some natural climate experiments, and I want to give you one example of one. That experiment is the eruption of Mount Pinatubo in June of 1991. The Pinatubo volcano put enough dust into the atmosphere to qualify it as the second biggest eruption of the 20th century. It put enough dust in the atmosphere, high into the atmosphere, to produce a measurable global cooling. When the Mount Pinatubo eruption went off, climate scientists said let's run our model with the amount of dust that this eruption has put into the atmosphere, and let's see what our model predicts for the climate of the next few years.

They ran their model, and they then waited a few years until we had global temperature records for the years following the Mount Pinatubo eruption, and they then compared them. I want to show you a graph of that result. The Mount Pinatubo eruption was really a natural climate experiment, and the graph I'm showing has a dark curve that represents the actual observed global temperatures in the years following the Pinatubo eruption. There are two dashed curves, which represent the results of model runs. They're slightly different because they make slightly different assumptions about the amount and properties of the dust the volcano ejected into the atmosphere, but they're pretty similar. You can see very clearly a good qualitative—and largely quantitative—agreement. As soon as the volcano erupted, the global temperature fell. It fell quite significantly, by about 0.5°C, not quite 0.5°C, and then it took a number of years to recover. By the way, if you look at Earth's global temperature in the early 1990s, or look at records of what were the warmest years on record, the years 1992, 1993, and 1994 tend not to be as high up in those lists as you think, and that's because of the Mount Pinatubo eruption causing a global cooling that was measurable and reproducible with climate models.

Let me summarize. We have simple to complicated models; the more complicated ones being large computer programs that run and try to simulate the climate, try to project the climate of the future. They include things like feedback effects. They include sub-grid parameterization effects for very small-scale phenomena. They are verified by a number of verification schemes, including particularly this natural Mount Pinatubo experiment, so we have some confidence that they give us a reasonable projection of what future climate ought to look like under conditions we specify.

Lecture Nine
Impacts of Climate Change

Scope:

A rise in the global average temperature is just one aspect of anthropogenic climate change. The temperature rise itself will not spread evenly over the globe; for example, it will generally be greater over land and at high latitudes. Extreme events, such as heat waves, intense precipitation, and droughts, will be more frequent, and storm intensities will likely increase. As ocean water heats, it expands, and this effect will raise sea level. Melting glaciers and land-based ice will augment sea-level rise. Although melting sea ice will have little direct effect on sea level, freshwater from such melting might disturb patterns of ocean circulation, such as in the Gulf Stream, possibly resulting in sudden "surprise" changes in climate. Absorption of anthropogenic carbon dioxide makes the oceans more acidic, putting shell-forming organisms at risk. Many of these changes have feedback effects that can further exacerbate the impact of climate change.

Outline

I. The most obvious effect of climate change is a rise in the global average temperature.

    A. Future climate change depends on human activities, such as greenhouse gas emissions and patterns of land use. For this reason, projections of future climate take the form of different scenarios based on different assumptions about human behavior.

        1. A set of scenarios used by the Intergovernmental Panel on Climate Change (IPCC) describes the future of human society on a two-dimensional space. One dimension is economic versus environmental emphasis; the other is globalization versus regional differences.

        2. The different scenarios give different projections of future greenhouse emissions and global temperature. The projections for different scenarios are obtained by averaging over projections of many different computer models. These projections suggest a 21st-century global temperature increase in the range of 1.5°C to 4.5°C. Statistical analyses show that there's little chance of the change being much lower than 1.5°C, but there's a non-negligible chance that it could be higher than 4.5°C. See Figure 15.

B. Why should a temperature rise of only a few degrees have significant effects?

    1. That few degrees is a *global average*. The rise will be more substantial in certain areas—particularly the polar regions and over almost all land areas.

    2. Lecture Three showed how only about 6°C separates the present-day climate from the depths of an ice age. Thus, a few degrees in global temperature is climatologically significant.

    3. Analysis of statistical distributions shows how even a small shift in the mean value of a quantity (such as temperature) makes a significant change in the likelihood of extreme events.

II. Extreme weather events, such as those listed below, will become more likely with increasing temperature.

    A. Heat waves. Although it's difficult to attribute any single event to anthropogenic climate change, statistical analysis of the 2003 summer heat wave in Europe that killed tens of thousands of people suggests that this event lies so far out on the normal distribution curve that there's at least a 50% chance it was caused by anthropogenic climate change.

    B. Intense precipitation events.

    C. Droughts.

    D. Intense tropical storms. Here, as often with climate change, a number of factors complicate the picture. Warming ocean waters should contribute to rising tropical storm intensity, and as warming penetrates deeper into the ocean, storms may become more robust because they're less likely to stir up cooler water from below. On the other hand, changes in wind patterns might shear apart nascent storms before they can become fully formed.

III. Sea-level rise is one of the most important long-term impacts of climate change. Measuring sea level is difficult because the sea surface isn't flat, because of tidal variations, and because the continental edges are themselves rising in some places and falling in others. Earlier measurements come from tide-gauging stations; modern measurements, from satellites.

A. During the last ice age, sea level was some 120 meters (about 400 feet) below its current level. The oceans rose rapidly as ice melted, but by 6000 years ago, the rate of rise had slowed to about 0.1 to 0.2 millimeters per year.

B. Sea-level rise today is caused primarily by two factors:

    1. Thermal expansion of the ocean waters as they warm.

    2. Melting of land-based glaciers and ice sheets. Melting of floating sea ice contributes almost nothing to sea-level rise, although it may have other effects.

    3. Other contributions to sea-level rise include melting permafrost and human alteration of the water cycle. "Mining" groundwater for human use adds to sea-level rise, while dams block natural flows and keep water from reaching the oceans.

    4. As usual, there are complicating factors. For example, increased precipitation may lead to more snow in polar regions. In the very cold Antarctic interior, this effect actually removes water from the oceans, slowing the increase in sea level.

C. The 20th century saw sea level rise at about 1–2 millimeters/year—some 10 times its rate in the past few millennia.

D. Projections suggest a global sea-level rise in the range of 6 to 18 inches by 2100 for the A1 balanced scenario.

    1. Sea-level rise will be greater if we continue with a fossil fuel–intensive economy.

    2. Sea-level rise has a long lag time; thus, even if greenhouse emissions stopped, we would be committed to another roughly half a meter (20 inches) over the next few centuries.

E. A rise of a foot or so doesn't sound like much, but this adds to already high tidal levels and storm surges.

1. Sea-level rise will have a significant impact in low-lying areas, such as barrier islands, Florida, and Bangladesh—which could lose some 10% of its land area.

2. Rising sea level forces saltwater into underground aquifers and marine estuaries, contaminating water supplies in coastal areas and damaging the "nurseries" for many marine species.

IV. The poleward advance of species ranges discussed in Lecture Two will accelerate.

    A. Some species, especially trees, may not be able to keep up with the changes. They, and ecosystems that depend on them, may cease to exist in many areas and some may even go extinct.

    B. Tropical species, including disease vectors, will spread into temperate regions. Such diseases as malaria, Lyme disease, West Nile virus, and Dengue fever may become more widespread.

V. Uptake of anthropogenic carbon dioxide in the oceans results in acidification of the ocean water. This may affect the survival of shell-forming marine plankton, which serves as the basis of marine food chains. Such effects will first occur at high latitudes, and then spread equatorward.

VI. "Surprise" events may occur if the climate system responds with nonlinear "tipping point" behavior.

    A. An example of nonlinear behavior is a light switch: Moving the lever gradually doesn't do anything for a while; then, the switch suddenly jumps to the "on" position and the light goes from off to full brightness.

    B. Catastrophic "surprise" events aren't considered likely during the 21st century, but their probability will rise considerably after the year 2100.

    C. One such nonlinear "surprise" could be the sudden slipping of a large land-based ice sheet into the sea.

        1. Possible causes include the lubrication of the ice-land interface by increased meltwater flowing below the ice and the melting or breakup of sea ice that helps keep land-based ice from sliding into the sea.

2. The West Antarctic ice sheet is of particular concern; this would raise sea level abruptly by some 3 meters, or about 10 feet.

D. A second nonlinear "surprise" would be a major upset in patterns of ocean circulation.

1. One possible cause would be the injection of freshwater into the northern North Atlantic as a result of the melting of sea and land ice. This effect is already weakening the so-called *thermohaline circulation* that transports warm surface water toward northern Europe.

2. An abrupt shutdown of the thermohaline circulation could, ironically, cause Europe to cool in response to global warming. However, images of a European deep freeze are probably unrealistic. Nonetheless, we still have a lot to learn about ocean circulation.

VII. Effects on human society and the global economy are beyond the scope of this science-based course. But one recent study suggests that climate change in the coming decades could reduce the global economy by some 20% unless major steps are taken to curb greenhouse emissions and other climate-changing activities. Acting now might cost just 1% of global economic output.

Suggested Reading:

IPCC 4, Summary for Policymakers.

Houghton, chapters 6–7.

Wolfson, chapter 15, section 15.2 to end.

Going Deeper:

IPCC 4, Technical Summary.

Stern (see also Web Sites to Visit, below).

Stott, Stone, and Allen.

Web Sites to Visit:

Pew Center on Global Climate Change, http://www.pewclimate.org/global-warming-in-depth. Both the Economics and Environmental Impacts links found on this page carry more detailed information relevant to this lecture.

British Treasury Department, http://www.hm-treasury.gov.uk/independent_reviews/stern_review_economics_climate_change/sternreview_index.cfm. From this site, you can download all or part of the 700-page Stern Review on the Economics of Climate Change.

Questions to Consider:

1. A friend asks how a rise of a few degrees in global temperature can possibly have a significant effect given that daily temperatures frequently vary by more than 10 times that much. How do you answer?

2. This lecture concentrated on physical, biological, and ecological effects of climate change. In what ways might these effects have an impact on human society?

Lecture Nine—Transcript
Impacts of Climate Change

Lecture Nine: "Impacts of Climate Change." What is climate change going to do to the world, and to human civilization? The most obvious effect of climate change is a rise in global temperature, but there are plenty of other impacts as well. This lecture is going to cover those impacts of climate change, starting with the rise in temperature itself.

There's an issue here. When I talk about future climate change, there's no single right answer to what the climate in the future will bring. That's because we don't know what human beings are going to do, and future climate, as we've now seen, depends on human activities, particularly how much greenhouse emissions we put into the atmosphere, and other effects as well. So how do we deal with this? We have to make up little scenarios, little storylines, to tell about what the future might look like in terms of human activities, and then we can use our climate models to calculate the climate that might result from those activities. I'm going to describe some scenarios developed by the Intergovernmental Panel on Climate Change, the IPCC.

Let me pause and just say a little bit about the IPCC because it's worth looking at what this organization does, and also understanding what this organization does not do. The IPCC is an outgrowth of the United Nations Environmental Program and the World Meteorological Organization. It consists of informal gatherings of hundreds of scientists, policymakers and others to talk about climate change and the state of climate science, and to make recommendations about what we might do about climate change. The IPCC is not a research organization. Its main job on a scientific side is to draw together from the open literature and make summary statements about the state of climate science.

The IPCC publishes many, many documents on different aspects of climate change. The most important ones are the assessment reports that are published roughly every six years. The fourth assessment report has just come out, and so that's the latest statement on summarizing maybe six years before of climate science. There are also reports on how to mitigate climate change, there are also reports on how to adapt to climate change, but the only report that I'm really concerned with in preparing this lecture series is the report on the science of climate change. There are plenty of other sources for the science of climate change besides the IPCC. The IPCC operates by a kind of consensus. They have to get approval from a large body of scientists and policymakers before they can publish the information they publish. If anything, it's kind of a least common denominator understanding of what the state of climate science is at any given time.

What the IPCC put together to envision the future, these so-called storylines, are scenarios that are based on kind of a two-dimensional arrangement of what human behavior might be like. They divide human nature in these two dimensions. One is whether human society decides to emphasize economic growth and economic development, and those are called the "A" scenarios. The letter A distinguishes them from the B scenarios, and the B scenarios envision a future society that puts more emphasis on environmental concerns, environmental preservation, and so on. The A scenarios are economic emphasis; the B scenarios are environmental emphasis. Then their scenarios are also numbered. The "1" scenarios assume that globalization—that is, the spread of economic cooperation, the spread of cultural norms, throughout the world will continue. That has some important implications because it means, for example, that there will be more universal education, particularly of women, and that turns out to be a very important aspect of affecting future population. A number of other globalization factors go into the 1 scenarios.

The "2" scenarios, on the other hand, assume that the world stays pretty much regionally divided, that different regions have their own subcultures, that economic cooperation doesn't develop to a very great extent. Those are regionally based scenarios, and then there are the global scenarios, the 1 scenarios. So crudely, this sort of two-dimensional space of scenarios for the future is divided into four scenarios—the A1 scenarios, which emphasize economic development in a globalizing world; the A2 scenarios, which emphasize economic development, but in a world that stays more regional; the B1 scenarios that talk about a world with environmental emphasis and globalization; and the B2 scenarios that talk about a world that remains regionally divided, but has environmental emphasis. Those are the scenarios.

Within that, the IPCC actually considered some 30 or more different sub-scenarios. The only ones which I'm going to talk about are within the A scenarios, the A1 scenario. There is the A1FI scenario, and that stands for fossil intensive. That's a scenario in which business as usual continues, and we continue to burn fossil fuels at an ever-greater rate, as we in fact are doing today. There's an A1B scenario—again, economic emphasis globalization—and the B scenario stands for balanced. We replace some of our fossil energy with non-fossil energy, non-carbon-based energy; we get a more balanced picture. And then there's the A1T, T for technology, in which new technologies replace much of fossil fuel technologies. Those create very different scenarios, as you can imagine, for future carbon emissions.

I should mention one other thing about these scenarios, with respect to population. In the globalization scenarios, the 1 scenarios—the A1 and B1 scenarios—the world population peaks at about the middle of the 21st century at somewhere around 9 billion people, and then it begins to decline. In the regionally based B2 and A2 scenarios, the population continues growing until the year 2100, the final year in which most of the projections are made. That has important implications for energy consumption and for the amount of greenhouse emissions. The globalization scenarios, partly because of the spread of education and economic development, and particularly the education of women, results in lower birthrates and a world population that peaks in 2050. That's a profound thing to think about. When I started teaching environmental studies, we always talked about the doubling time for the world population. We don't talk about that anymore because the world population is unlikely ever to double again. It would be a wonderful thing in many ways if it were to peak in the middle of the century. That's what the A1 and B1 scenarios have it doing.

What happens with these scenarios? The IPCC has worked out detailed assumptions under each of these scenarios and sub-scenarios for emissions of all kinds of different greenhouse gases, for all kinds of other factors that affect climate. You can look at tables, and tables, and tables of numbers they've developed that are their assumptions in each of these scenarios for what will happen with greenhouse emissions. When we put that all together and put it into climate models, what's happened is a number of climate groups around the world have run their climate models with these different assumptions. Different models, as you learned in the last lecture, give slightly different results, so we've averaged together the results for these different model runs and come up with projections for what future greenhouse emissions, particularly CO_2 emissions, might look like under these different scenarios, and for what the resulting temperature increase might be.

Let's take a look at some of those graphs. We have two graphs: one about CO_2 emissions, and the other about projected temperatures. The CO_2 emissions projections differ dramatically. If you look at the graph on the left side, the dotted line is the A1FI scenario, the fossil-intensive scenario. You see, under that scenario, carbon emissions rising dramatically. They're starting at the left at that seven Gt/year that I mentioned—these are in Gt of carbon per year—7 Gt/per year the same flow of carbon that I mentioned is coming from the fossil fuels and going into the atmosphere. They rise dramatically under several of these scenarios, particularly under the A1FI scenario. Under the A2 scenario, they also rise dramatically. In some of the scenarios, the B2 scenario and, to a lesser extent, the B1 scenario, they rise for a while and then decline. That makes a very big difference by the year 2100 in how much CO_2 we have put into the atmosphere.

On the right are the temperature projections. They show less dramatic variation, and that's because they depend on the accumulated CO_2, some of which is already there. Under all these scenarios, more carbon is still being put into the atmosphere, even if the rate of emission goes down. The temperature range projected is not nearly as great as the range of projected greenhouse emissions by the end of the 21st century. Let's take a look at the global temperature. There's a light band that marks the range in which most of these models' predictions lie. The A1FI scenario, not surprisingly, shows the greatest temperature increase; it's a temperature increase here of about maybe 4.5°C over the century to come. The lowest rising scenarios include the B1 scenario. The B1 curve shows a rise in temperature of perhaps only 1.5°C. Somewhere in that range, 1.5 to 4.5°C, is probably what we can expect for a temperature rise over the course of the 21st century. That doesn't sound like a whole lot. I want to take a little bit of time to describe why those numbers mean something important.

Let's just pause a second and look again at these graphs. We have a choice here. These are scenarios based on assumptions about human behavior, about what we decide to do about the rate of our greenhouse emissions. If we continue business as usual with the A1FI scenario, we will end up, in the year 2100, with greenhouse emissions that exceed 25—that approach almost 30—Gt/year. Right now it's about seven Gt/year. Incidentally, the A2 scenario actually exceeds the A1FI scenario by the end of the century in greenhouse emissions, and that's because that growing population actually causes more greenhouse gases to be emitted right at the very end; although through most of the century, the A1FI scenario results in the most greenhouse emissions, but very end, the other one catches up because it's the one that has the growing population, whereas the A1FI scenario has a population that peaks in 2050.

The bottom line from these CO_2 and temperature projections from many, many computer models—ensembles of different computer climate models, being run with these assumptions from the IPCC scenarios—is that we can expect a temperature rise of somewhere in the range of 1.5 to 4.5°C over the coming century. By the way, statistical analysis of the distribution of probable temperatures suggests it's very unlikely to be less than 1.5°C, but there's a substantial chance that it might be greater than the 4.5. It's not kind of symmetric in that range— very unlikely to be less than the bottom end at 1.5°C; a substantial probability that it might be significantly greater, 6° or 7° even, than the uppermost of those model run projections.

Why should a temperature rise of a few degrees matter? I raised that question earlier, way back in Lecture One. I answered it in several ways, and I'm going to answer it in more detail now. First of all that rise, remember is the global average temperature, averaged over the entire planet, over the Arctic, over the equatorial regions, over the oceans, over land. The rise will be greater, by roughly a factor of two, on all land areas. Not every point in land will go up; some points will still go down, but on average, the rise in the temperature over land will be bigger than the global temperature rise by very roughly, roughly, perhaps a factor of two. The polar regions will rise much more dramatically. From what you learned in the last lecture about the ice-albedo feedback, you can kind of see why that's going to occur.

In lecture three, I gave you a number that I wanted you to walk away from this course with, and that was the number that represented the temperature difference between the present temperature, the present interglacial warm spell, and sort of the average temperature in an ice age. It was on the order of 6°C. When somebody tells you the global temperature in the next century might go up 4°C or 3°C or whatever, that's not much smaller than the difference between now and an ice age; a difference that meant, again, northern Europe and North America being covered by a few miles of ice, a couple miles of ice. That's a big climatic difference. Globally, a difference of a few degrees can be huge in terms of the global climate. That's another reason.

Finally, a third reason is that, if you look at statistical distributions in general, even a small shift in the mean value of some statistically distributed quantity can result in big changes in the extreme values. That's another reason why a few degrees makes a big difference. If you think of something like the grades on a test in a course, or the temperature extremes, the maximum temperatures you get on a given day, they tend to distribute themselves in a bell curve. There are a lot of values grouping right near the peak of the curve, right near the average; very few at the tails of the distribution, the extremes that are very far from the average are very, very unlikely, and that's what makes this bell-shaped distribution. If the average temperature goes up a little bit, for example, the distribution of temperatures moves only slightly; let's say it goes up a degree or something. The average value goes up a degree, and that whole bell curve moves over only slightly.

That doesn't very much affect things that are happening around the mean. But what it does do is affect much more dramatically what's happening at the edge. If you work this out mathematically, way out in the tails of that distribution, in the very rare events that are very far from the mean, even a very small shift in the mean value dramatically raises the likelihood of those events. If you've seen my Teaching Company course *Physics in Your Life*, I describe the same phenomena with respect to food spoilage. Why is it that you can leave a bottle of milk in the refrigerator for two weeks, and you take it out on the counter, and it's spoiled in an hour? In the grand scheme of things, in absolute temperatures, that difference isn't all that much. Why? It's because the extreme events, the high-energy molecules in that case that cause the chemical reactions that cause spoilage, and the biological activity that causes spoilage, are much more enhanced with even a few degrees of temperature increase.

The same is true in climate. Any extreme events that occur in climate, even though they're very unlikely, their likelihood is greatly enhanced by even a slight shift in the mean temperature. One thing we should expect, in addition to just the temperature going up, is an increase in extreme—unlikely, but extreme—events. One such event is heat waves. I mentioned heat waves before. Heat waves are extreme events where the temperature remains very high for a long time. I'll just give you a little anecdote from today because I heard the weather report this morning, and today in the Washington, DC area, it's going to be the 25th day in a row in which the high temperature is above average.

If you think about that, on average, the high temperature should be above average half the time and below average roughly half the time. This is 25 straight days in a row in which the high temperature has been above average. That's an extreme kind of event. That's why the meteorologists took note; that's an unusual event. You might get a string of two or three days, maybe a week, where the high temperature is above average; that wouldn't be too unusual. But 25 days in a row? There's something unusual about that. That lies way out on the extreme of that bell curve. It's that kind of event—that kind of extreme unlikely event—that's enhanced by global climate change, by even a slight change in the mean temperature.

You may remember that in the early 2000s, there was a heat wave in Europe that killed tens of thousands of people. You can't attribute any single event, like the 25-day streak I'm just talking about, or that heat wave in Europe, to human-caused global warming. You can't pin down any one event, any more than you can say one particular case of cancer was caused by radiation or something. However, statistical analysis of that heat wave in Europe says that was so far out on the extreme, on the tail of the bell curve, that there's only a 50% chance that it was caused by natural effects. There's a 50/50 chance that that heat wave that killed tens of thousands of people in Europe was anthropogenic in origin, caused by anthropogenic climate change.

There are plenty of other extreme events that are likely to occur more frequently because of a few degrees change in the average temperature. They include intense precipitation events, they include droughts; they include intense tropical storms. I talked about tropical storms before and showed you a correlation between tropical storm intensity and sea-surface temperature. That, like many other aspects of climate, is a little bit subtle in nuanced. For example, we know that warming ocean waters should contribute more energy to make the storms go. As the warming penetrates deeper into the ocean, the agitation of the storm, which in previous eras would have brought up cool water that might have quenched the storm, will now bring up warm water. That may make the storms more robust and longer lasting. On the other hand, there are subtle changes in wind pattern. Hurricanes, when they try to form, if there's too much wind shearing along high altitudes, that will simply shear the top of the hurricane off, and the hurricane won't be able to form. There are subtle effects here, but we do expect more intense tropical storms.

One of the most significant and most talked about aspects of climate change is sea level rise. That's not an easy thing to measure, sea level itself. You might think well, there's a uniform sea level all around the world, but that's not true. The sea level varies from place to place. Furthermore, the land areas are going up and down, still. One reason they're going up and down is they're still rebounding from the last ice age, when huge blocks of ice weighed the continents down. When the ice melted, they began to rebound, and that rebound is still going on, 10,000 years after the ice melted. So, it's difficult to measure sea level. In the past, we used tide gauges to measure sea level, gauges at particular locations that measured the sea level as the tide came in and out. But even their measurements, again, have to be corrected because the land on which they're sitting may itself be moving up and down. More recently, we use measurements from satellites that look down on the ocean, and can get measurements over broad areas of the sea surface, so we have a better feel for what's going on.

During the last ice age, sea level was 120 meters roughly—that's about 400 feet—below what it is now. Sea level changes dramatically with time. You can imagine what that would do to the world's coastal cities if sea level suddenly dropped 400 feet, which it did because all that water was locked up in ice. The sea level rose quite rapidly as the ice melted, but by about 6,000 years ago, that rate had slowed to about a 0.1 millimeter per year. That's not very much, 0.1 to 0.2 millimeter per year, a tiny, tiny little rise every year.

Today, we see a significantly greater sea level rise of about 10 times that, about 1 to 2 millimeters per year.

What causes sea level rise? You probably think it's caused by all that melting ice that I talked about earlier. That's some of the cause, but the most significant cause is simply the thermal expansion of the water. The ocean acts like a giant thermometer. The water expands when it's heated because the molecules get farther apart because it got more energy, and consequently, the whole water level rises. By the way, that's a process that will take some time to complete, as the heat penetrates further and further into the ocean. It's only the surface layers that have been affected so far. Thermal expansion of the ocean water causes sea level rise.

The melting of land-based glaciers and ice sheets causes sea level rise, but the melting of floating ice causes essentially no sea level rise. You can find that out for yourself by putting a few ice cubes in a full glass of water, dry it all off on the outside, let the ice cubes melt, and you'll see there's no change in the water level, and that's because the density of ice is less than the density of water by just the right proportion to make that work out. That's how floating things work. Land-based ice melting causes sea level rise. Floating ice does not cause sea level rise, although it does cause other changes that I'll mention. There are some other contributions. For example, melting of permafrost probably causes sea level rise. We humans alter the water cycle in ways that are actually quite significant when we mine water out of underground aquifers. Mining is really the right word for that non-renewable resource.

When we mine water from underground aquifers and let it flow through surface waters to the ocean, that contributes to a rise in sea level. When we build dams that block water and keep it from going to the sea, and it gets evaporated instead, there we lower sea level. Those effects are small, but they're significant. Again, this is a complicated subject. For example, Antarctica—I mentioned earlier its interior is a cold, dry desert. Any precipitation that falls on Antarctica is locked up as ice for a long time. It would take a really large temperature change to start melting anything in the interior of Antarctica. Therefore, if global warming causes more precipitation, one effect will be to remove water from the oceans and deposit it onto central Antarctica. That's not a huge effect, but it's enough to counter some of the increasing sea level caused by some of these other effects I've mentioned. Again, the climate system is subtle.

The 20th century, as I said, saw a rise in sea level of about 1 to 2 millimeters per year. That's about 0.5 to 1 inch every decade. It doesn't sound like a whole lot. If you look at the projections for sea level rise that come out of our best climate models, they now predict a rise of some 6 to 18 inches by the year 2100, in the A1 balance scenario—it would be more in the A1FI scenario—it would be less in the A1 technology scenario, and it would be different in the other different scenarios. Six to 18 inches doesn't sound like a lot, and you may have heard tales of very large sea level rises, and how we're going to inundate the coastal cities. This is one projection in climate science that has actually become moderated over time as our climate models have gotten better. We used to think it might be more like half a meter, about 20 inches, in the coming century. Now it looks like it may be a little bit less than that. It will be greater if we continue business as usual, with our fossil-intensive economy.

I mentioned also that sea level rise is going to go on for a long time because it takes a long time for ice to melt, and because it takes a long time for the heat to penetrate and expand the deeper waters. Even if we stopped greenhouse emissions right now we've probably committed ourselves to about 20 inches, so maybe almost two feet, of sea level rise over the next few centuries. That doesn't sound like much, but it already adds to high storm surges and high tides. If storms get more intense, that coordinates with the more intense storms and the higher sea level to make a double whammy in terms of the storm surges.

There are some other effects also. Very low-lying areas like Florida or Bangladesh will see significant loss of land, even from a very small amount of sea level rise. Bangladesh still stands to lose maybe 10% of its land area in the coming century from sea level rise. Rising sea level also forces salt water into freshwater aquifers, and it also contaminates marine estuaries by altering the salt balance. That damages the nurseries for much of marine life. There's some other problems that we have to worry about from sea level rise. It's not the dramatic flooding you may be thinking about, but it has some subtle effects, and it will be significant. That's only until the end of the 21st century that I'm talking about. Things may get worse after that.

Another thing I discussed in Lecture Two was this poleward advance of species that we measure as an indication of climate change. That will certainly continue. But some species, particularly things like trees that can't move rapidly to keep up with changing climate—they can only seed a little more at the northward edge of their range and so on—they may not be able to move fast enough, and they may go extinct, at least in some regions. Then other species that depend on those trees may disappear also. Tropical species, particularly disease vectors like mosquitoes and so on, may move into northern climates, or southern climates in the Southern Hemisphere. Malaria, Lyme Disease, the tick-born disease, West Nile Virus, Dengue Fever; disease like that may become much more widespread in a warming world.

There's another effect that recent scientists are only recently beginning to worry about; it's really beginning to get people's attention. This is something that is not a direct result of global warming, but it is a direct result of anthropogenic CO_2 emissions. I've mentioned that about half the CO_2 we put into the atmosphere doesn't stay there but ends up in the soils and the oceans. That which ends up in the oceans changes the acidity of the ocean waters because CO_2 dissolves in water to make a mild acid. That's why soda pop is not so good for your teeth, because the soda water is acidic, and it can rot your teeth. What's happening with the acidifying oceans is that the ability of shell-forming organisms to form their shells and keep their shells together goes down as the water becomes more acid. Another subtle effect of our human greenhouse gas emissions, not related directly to the warming that those emissions cause, is when they are taken up in the oceans, the oceans are acidifying. That may have a significant impact on the marine ecosystem.

Finally, let me end by mentioning the so-called surprise events that may occur, and may be quite dramatic, if global warming reaches a certain point. Most of the phenomena we're talking about so far are sort of proportionate to cause. If the temperature increase doubles, sea level rise might double, and so on. But there may be surprise events, so-called "tipping point events," that switch on suddenly, despite only a small change in temperature. Let me give you an example of something that is a so-called non-linear or tipping point response, and that's a light switch. Take an ordinary light switch and start pushing on it and nothing happens. The light doesn't get a little bit brighter and a little bit brighter. No, it stays off. Then suddenly the switch flips into the on position, and the light comes on at full brightness. That's a very non-linear effect, a very small change from just before it was going to go on, to when it goes on fully. A very small change in the position of the switch makes a huge change in the intensity of the light. There might be things like that lurking in the climate system, surprises like that.

I need to say that most climate scientists do not consider such surprises likely in the 21st century. They will become more likely as things progress, but in the present century they aren't considered very likely. On the other hand, if they occurred, they could be catastrophic. One such surprise event would be the slippage of a large land-based ice sheet into the ocean. I mentioned before that one worry we have about melting of land ice is that the melt water running under the ice sheets lubricates the ice sheets. If they're unstable, they could suddenly fall into the oceans and slide into the oceans and cause a large rise in sea level. Another reason that might occur is that certain sea ice at the edge of the land area is actually holding some of that land ice on continental areas like Antarctica. If the sea ice melts, then that land ice may be free to slide in. Of particular concern is the west Antarctic ice sheet, a large ice sheet that appears to be somewhat unstably positioned on the continent of Antarctica. If it fell into the water, it would raise sea level quite abruptly by some three meters, which is about 10 feet. That would clearly be a problem for us.

A second non-linear surprise would be an upset in the global circulation of the ocean waters, which is driven by changes in temperature, and also by changes in salinity. The particular concern here is in the north Atlantic, where melting Arctic sea ice will cause an influx of freshwater—because when ice freezes, it leaves the salt behind—and that will cause changes in the circulation pattern. The particular worry there is that the circulation that carries warm water from the Caribbean and the southeastern United States area up to Europe, and keeps Europe's climate quite mild—the Gulf Stream— that that might slow down. Despite what you've seen in the Hollywood movies, a freezing of Europe is unlikely. In fact, this effect, even if it occurs, is only likely to reduce the amount of global warming Europe would otherwise experience. But it's kind of a worrisome thing to imagine shutting off this enormous circulation

©2007 The Teaching Company

pattern that affects the entire ocean. That's one of the concerns people have.

Let me end with a comment that I really is outside the scope of this course, and that's a comment about the effects on human society and the global economy. Again, this is a course about science, and this is not an area I want to go into, but I do want to make two observations. A lot of people don't want to think about global warming because it's going to cost us money to solve the problem. It's going to cost us money to develop new energy technologies, for example, and we're worried about the effect on the economy. A new report suggests that if we don't do something about global warming, that could decrease the global economy by some 20%; whereas it might cost us only about 1% every year to fix the problem now.

One other thing I think anybody who worries about global climate change should think about is this. Countries like the United States are better positioned to adapt to the changing climate—they're also largely the countries that are causing the problem. One thing this is going to do is exacerbate the gap between rich and poor nations. In a world that is already quite hostile and combative, I think we need to think about whether that's something we would like to have happen. But that's really beyond the scope of this course, which is a course largely about science.

Lecture Ten
Energy and Climate

Scope:

Humankind's thirst for energy is the dominant reason for our increasing influence on Earth's climate. The world uses energy at the prolific rate of about 15 trillion watts. In North America, per capita energy consumption is more than 100 times our own bodies' energy output. We use energy for a range of purposes that enhance our quality of life, from keeping warm to providing mobility to powering industry. Nearly all of humankind's energy comes from the fossil fuels coal, oil, and natural gas. Burning these fuels naturally releases carbon dioxide, which is not a pollutant in the traditional sense but, rather, an expected and, indeed, desired product of fossil-fuel combustion. The result is the 7 gigatonnes of fossil carbon released into the atmosphere annually, as described in Lecture Seven. The high living standards of industrialized countries are closely related to energy consumption.

Outline

I. Humanity uses energy at the rate of 15 trillion watts—the equivalent of 150 billion 100-watt light bulbs.

    A. The *watt* measures *power*—the *rate* of energy consumption or generation. That 15 trillion watts can be expressed in many equivalent ways:

        1. In English units, that's nearly 500 quadrillion BTUs per year.

        2. It's the energy equivalent of some 80 billion barrels of oil per year (actual oil consumption is about 30 billion barrels per year), or more than 2500 barrels of oil every second.

        3. For scientists, it's about 5×10^{20} joules every year (500 EJ/year).

    B. Per capita energy consumption is correspondingly large.

        1. With world population nearing 7 billion, that 15 trillion watts is more than 2000 watts per person.

        2. By metabolizing food, the average human body produces energy at the rate of about 100 watts. On average, then, each human uses energy at about 20 times his or her own body's energy output.

3. Energy consumption is far from uniform over the globe. In the most energy-intensive countries, it exceeds 10,000 watts (10 kW) per capita, the equivalent of more than 100 human bodies' worth of energy.

II. What do we do with all this energy? Four sectors—industrial, transportation, residential, and commercial—account for nearly all our energy use.

A. In the United States, where we have about 100 "energy servants" each, you can think of the percentage of energy use in each sector as the number of servants in that sector.

B. Some 28 energy servants are at work round the clock to transport you and the goods you use.

C. Another 32 labor on your behalf in industry.

D. Some 22 are at work in your home, keeping you warm or cool; making hot water; cooking; and running lights, entertainment systems, and computers.

E. The commercial sector employs some 18 energy servants, keeping on the lights, copiers, and computers in offices, hospitals, and educational institutions and running the lights, freezers, and cash registers in our supermarkets and stores.

III. Who are our energy servants? They're the mix of sources that we use to produce energy. See Figure 16.

A. Some are fuels, substances that store energy. These include fossil and nuclear fuels, as well as fuels made from biological materials.

B. Others are tapped from natural flows of energy. These include the energy of flowing water, wind, and tides and currents, and the nearly steady flow of energy from the Sun.

C. Expect much more on fuels and energy flows in the next lecture.

D. The mix of energy sources differs from country to country, depending on the country's resources or its wealth and, therefore, its access to imported energy.

1. In both the United States and the world as a whole, the fossil fuels coal, oil, and natural gas provide the overwhelming contribution to the energy supply, approaching 90% in both cases.

2. Although this pattern holds throughout most of the world, there are exceptions. For example, France gets a significant fraction of its energy from nuclear sources; Norway, Congo, and Brazil, from water power; and Iceland, from geothermal sources.

E. The climate connection: Fossil-fuel combustion and other fossil energy–related activities are responsible for some 83% of U.S. greenhouse gas emissions.

1. CO_2 from combustion of fuels dominates the greenhouse emissions related to fossil fuels; other energy-related sources include methane leaks from natural gas systems, refinery emissions, and methane from active and abandoned coal mines.

2. Significant sources not related to energy production include methane from landfills and agriculture, carbon dioxide from cement production, methane from sewage treatment, and many industrial processes.

3. Included in these figures are the effects of land-use "sinks" that remove CO_2 from the atmosphere.

IV. What does our prolific energy consumption buy us? (We're moving out of pure science but staying quantitative!)

A. For many countries, there's a direct correlation between energy consumption and material well-being, as measured by the gross domestic product (GDP). See Figure 13.

B. But some countries are less efficient in their use of energy, producing lower GDP for a given rate of energy consumption. Others are more efficient.

C. Overall, a sample of many countries shows a general correlation between energy consumption and GDP.

D. But some economists question the GDP's appropriateness as a measure of human well-being. An alternative, the United Nations' Human Development Index (HDI), shows an initial rise with increasing energy consumption but then a leveling off, after which more energy consumption does not increase a country's HDI.

Suggested Reading:
Smil 2006, chapters 1–4.
Wolfson, chapter 2.

Going Deeper:
Smil 2003, chapters 1–2.

Web Sites to Visit:
U.S. Department of Energy, Energy Information Administration, *Annual Energy Review*, http://www.eia.doe.gov/emeu/aer/. Published yearly and made available on the Web, the *Annual Energy Review* is a comprehensive compilation of energy statistics for the United States. There's a less detailed section entitled "International Energy." Most of the data are displayed in both tables and graphs, and the tables are available for download as spreadsheets.

U.S. Department of Energy, Energy Information Administration, *International Energy Annual*, http://www.eia.doe.gov/iea/. A more comprehensive compilation of international energy statistics.

International Energy Agency (IEA), http://www.iea.org/Textbase/stats/index.asp. This site lets you generate energy data for your choice of country, region, and energy type from the IEA's vast database. A click on the Key Statistics link gets you to a free download of the annual *Key World Energy Statistics*.

U.S. Environmental Protection Agency (EPA), "Greenhouse Gas Emissions," http://www.epa.gov/climatechange/emissions/index.html. This EPA site is an authoritative source for detailed statistics on U.S. greenhouse gas emissions.

Questions to Consider:
1. If you're a resident of North America, you've got something more than 100 energy servants working for you, round the clock. Roughly how many are "working" in (a) transportation, (b) industry, (c) commerce, (d) your home?

2. Energy consumption is the dominant cause of anthropogenic

climate change. What are some others?

3. Discuss how a country's energy consumption relates to its standard of living.

4. Find out where most of your electrical energy comes from. This will vary from country to country and from state to state within the United States and may even depend on your particular municipality or power company.

Lecture Ten—Transcript
Energy and Climate

Welcome to Lecture Ten: "Energy and Climate." In the first six lectures, I laid the groundwork for a scientific understanding of how planetary climates are established, and particularly how Earth's climate is established, ending up with the cycling of carbon, that all-important element that is establishing the greenhouse effect in Earth's atmosphere. The next three lectures, which we've just completed, looked at humankind's contribution to the global carbon cycle and to other properties of the atmosphere that result in greenhouse warming. We've seen the science of climate. We've seen how human beings are altering the climate, and understand that from a scientific point of view. In the last three lectures, we look at how human behavior might be changed in a way that could avoid some at least of the global warming we anticipate.

Again, I'm not making prescriptions here for what I think human behavior should be. I'm simply making observations about what we are doing now and how those practices might change if we want to avoid at least some of the global warming. Some of it we can't avoid because we're already committed to it with the CO_2 we've already put in the atmosphere. Why is this lecture called "Energy and Climate"? Because of all the things we human beings do that have the possibility to alter climate, by far the most significant is our consumption of energy. In this lecture, I want to talk a little bit about how much energy people use, where that energy comes from, and what we get from that energy. Then in subsequent lectures, we'll look at what we might do differently.

How much energy do we use? That's a common question, but it's also the wrong question. It's the wrong question because the issue isn't how much energy we use, but how much energy we use in a given time, at what rate we use energy. How much energy do we use in a year? How much on average do we use every second? How much do we use every hour? We're really asking a question about the rate of energy consumption. It's very common to confuse energy—a stuff, a substance, if you will—with the rate at which it's being used, or produced, or whatever. That rate is called power, and that power is measured in watts. A watt is a rate of energy use. You know what a watt is because you know what a 100-watt light bulb is, or 1,000-watt hairdryer, or a 1,500-watt stove burner, or a 60-watt light bulb or whatever. You have a vague feel for what a watt is. I'm going to quantify that feel quite solidly in the course of this lecture.

I want to emphasize that a watt is a unit of power. It's an amount of energy per time. In particular, in the metric system, the unit of energy is the joule, named after James Joule; and a watt is a joule/second. That doesn't mean much right now, but I'll try to give you a better feel for what it means. Let me rephrase the question. At what rate does humankind use energy? Again, I'm phrasing that question because we're concerned, in this lecture, with how human beings cause greenhouse warming, and the dominant reason we do that is because we consume lots of energy. At what rate do we consume that energy?

We use energy at the rate of about 15 trillion watts—about 15 trillion watts—that's the equivalent of 150 billion 100-watt light bulbs burning. Now don't ask me: Is that 15 trillion watts every year or 15 trillion watts every hour, or what? It's not any of those things because it's already a rate. The word watt has that rate built into it. That 15 trillion watts is 15 trillion joules of energy every second. You can figure out by multiplying by the number of seconds in a year—which is approximately $\pi \times 10^7$, by the way—to figure out how much we use in a year, or in any other time interval you want but that 15 trillion watts is a rate at which we are using energy on average, around the clock, all the time, the entire species; 15 trillion watts, the equivalent of 150 billion 100-watt light bulbs.

Some other equivalents; there are lots and lots of units used to measure energy, a bewildering array of them. Different countries and different specializations use different energy units, and you just have to get used to there being several. Let me express that in several other ways. Frequently, energy is expressed in quadrillions of British thermal units (BTUs). The British thermal unit is a unit of energy; it happens to be about 1,000 joules, not quite; perhaps 1,055 joules. We express large-scale energy uses, like of whole countries or of the world, in terms of British thermal units/year. Again, we're talking about a rate.

That 15 trillion watts is equivalent to about 500 quadrillion BTUs/year. They're often called quads, 500 quads/year. The quad is a common unit you'll hear used for energy; a quad is a quadrillion BTUs. A quad is equivalent to 500 quadrillion BTUs, 15 trillion watts; 500 quadrillion BTUs per year they're equivalent to about 80

©2007 The Teaching Company

billion barrels of oil per year. If we got all our energy from oil—which we don't—we would be consuming about 80 billion barrels of oil a year, about 2,500 barrels of oil every second. One: If we got all our energy from oil and there goes 2,500 barrels up in smoke. That's the rate at which we would consume oil if all our energy came from oil. In fact, we consume about 30 billion barrels of oil a year, and that tells you that our oil consumption is a significant fraction; not quite half, but a significant fraction of the energy we consume is in fact coming from oil.

If you're a scientist—or if you like scientific notation—that 500 quadrillion BTUs/year, that 80 billion barrels of oil equivalent/year, that 15 trillion watts—not per year because the watt has got the time unit built into it; the others are expressed as an energy per time—is also equivalent to 5×10^{20} joules/year; or 500 exajoules/year, if I can use the "exa" prefix, which means 10^{18}. It's about 5×10^{20} joules, or about 500 exajoules/year, 80 billion barrels of oil equivalent per year, 500 quadrillion BTUs/year, or simply 15 trillion watts. Those are all ways of saying the same thing, namely the colossal rate at which humankind consumes energy. That energy is being consumed by roughly 7 billion people on the planet. So the energy per person, the energy per capita, that we're using is a considerably smaller number, but it's still big.

That 15 trillion watts, divided among a world population of 7 billion, comes out to about 2,000 watts per person. Don't ask me do I consume 2,000 watts every day or every year or what because, again, that's a rate. You, on average, as a citizen of the world, consume 2,000 watts. That's the rate at which you consume energy, around the clock, all the time. I suspect most of the people in my audience consume far more than that—and I'll get to that in just a moment—but let me focus on that 2,000 watts a minute because I want to look at that and compare it with the energies that our own bodies can produce. The human body, metabolizing its food, produces energy at the rate of roughly 100 watts. We're roughly equivalent to a 100-watt light bulb.

I want to give you a real feel for what that 100-watt number means. So what I want to do is do a little exercise here that will get you to feel, literally in your muscles, what 100 watts means. If you've seen my course *Physics in Your Life*, I had a hand crank generator there, and I got a Teaching Company employee to crank out 100 watts and light a 100-watt light bulb. That Teaching Company employee felt exactly what 100 watts felt like. We're not going to do that. We're just going to use our bodies to produce energy at the rate of about 100 watts. I'm going to do some deep knee bends. If you're not driving in your car, and you're sitting leisurely in front of your TV, and are suitably able, why don't you get up and do some deep knee bends with me, like this.

We're going to go down about half a meter, about 18–20 inches, and we're going to do that about once a second. One, two, three, four, up, down, up, down, up, down, up, down. I could keep doing that for a while, but I'd get pretty tired. My body is producing energy at the rate of about 100 watts. I'm not going to calculate that for you, but if you had high school physics and know about kinetic and potential energy, you could probably do that calculation and come to the conclusion that we're producing energy at the rate of about 100 watts. If you're a lot smaller than me, it's probably a little bit less. If you're a lot bigger than I am, it's probably a bit more, but it's somewhere in the range of 50 to 200 watts, or something like that, that you're producing if you're doing those deep knee bends.

I just argued that, per capita, the world population uses, on average, 2,000 watts—uses energy at the rate of 2,000 watts. What that means is that the average human being on this planet is consuming energy beyond that produced by their own body; in fact, about 20 times as much as produced by their own body. If you wanted to think about sort of energy servants that were working for you, doing the work of all that energy, every person on the globe has approximately 20 energy servants; that is, 20 human bodies' equivalent worth of energy production going on their behalf. That energy consumption is far from uniform over the globe. In the most energy-intensive countries, the energy consumption rate per capita exceeds 10,000 watts. There's 100 hundreds in 10,000, so that means in the most energy-intensive countries—of which the United States is one—the energy consumption rate is equivalent to somewhat over 100 human bodies' worth of energy per capita. You've got 100 energy servants working for you if you're in the United States.

Since the average is about 20, that means there are countries that have far fewer than that 20, so it varies enormously. For example, in Australia, the average number of energy servants, if you will, is on the order of about 70 to 80. That corresponds to something like 7,000 to 8,000 watts, seven to eight kilowatts. In the United States, it's about 11 kilowatts, actually about 110 energy servants. In Saudi Arabia, it's pretty high; it's almost 80 energy servants. In Japan, it's about 50. In Poland, it's about 35 or so. It varies around the globe. If you're a resident of Europe, it's probably about half what it is in the United States; that is, you use energy at about half the rate an American does if you're a European, on average. In some parts of the world, in Egypt, it's about 10 energy servants, about 1,000 watts. In Congo, it's just a few energy servants, just a little bit more than the human body can produce. It varies dramatically. In Switzerland, it's just about half what it is in the United States, and so on. So that number varies dramatically.

What I'd like you to remember, especially if you're in United States, or North America for that matter—Canada is actually just a hair higher—is that the average North American has a body, which is 100 watts, worth of energy, and has 100 human bodies worth of energy being produced in your name to supply you with all the things energy does for you, from transportation to refrigeration, to heating your house, to whatever. We'll look at that in just a minute. Picture this whenever you think about energy consumption. Picture yourself, your puny little body, and then 100 other human beings working just for you, with the energy output of their bodies, to supply all the other energy you use. Of course, that energy isn't coming from human bodies. There are certainly not enough people in the world to supply that energy if it were. But that's the equivalent in human bodies of the energy you consume.

What do we do with all that energy? It's conventional to sort of think about four sectors—industry, transportation, residential and commercial—as accounting for nearly all that energy use, and to sort of divide the energy use up into those sectors. In the United States, where we have about 100 of these "energy servants" for every person, you can think of the percentages as being the number of servants. In the United States, something like 28 of your energy servants—that is, 28% of the energy you use—they're working round the clock just to transport you and the goods you use. All those trucks whirring down the highway, all those airplanes delivering Federal Express packages or carrying you to the other side of the continent or whatever; if you add up all the energy they consume and divide by the population of the United States, you will find that about 28 energy servants, about 28 human bodies' worth, are supplying you with transportation.

Another 32 of them, the biggest single chunk, are working in industry on your behalf, producing the goods that you use. Neither of those energy servants are ones we see very obviously. I guess we see our transportation ones when we're powering our cars, but not a lot of the trucks that are delivering the goods and so on. We certainly don't notice every day the industrial ones. But most of the energy servants who are working for you—more than half the energy you consume, in other words—is in the transportation and industrial sectors. About 18% are in the commercial sector. What are they doing? They are keeping the lights on in your office buildings. They're running computers. They're keeping the hospitals operating. They're running the banks of freezers—some of them open to the surrounding air, very inefficient, in our supermarkets. They're keeping the lights on in the parking lots of the stores that are open until 10 pm and so on. They're doing all the things that need energy in the commercial sector.

About 22 of those energy servants, about 22% of your energy, are at work in your home. The dominant thing they're doing is heating or cooling your house. They're making hot water, as the second most energy-consuming thing they're doing. They're running your computers, your alarm systems; they're running your stoves, electric or gas. They are running your water pump; they're lights; they're running entertainment systems. They're doing all the things you do in your house. About 22 of your servants are at work in your house. Imagine having to cram 22 extra people into your house for every person in your house, in your family. Maybe there are four people in your family; that's 88 energy servants you've got to have in your house to keep that energy going.

That's your energy servants. Again, for residents of North America, you've got about 100 of them if you think of equivalent numbers of human bodies. If you want to think a little more scientifically, you're using energy at a rate of slightly over 10,000 watts, 10 kilowatts—again, that's round the clock. Those 100 energy servants, by the way, if they worked eight-hour shifts, you'd have to employ 300 of them, each working 8-hour shifts, because they're working round the clock for you. That 10,000 watts is a rate of energy consumption, and that's the average rate at which you consume energy round the clock.

Of course, sometimes you're consuming it a higher rate, sometimes at a lower rate. If you step in your SUV and turn the key, I guarantee you you're consuming energy at a much higher rate, but just for the relatively short time that you spend in your SUV.

Who are these energy servants? They aren't really people cranking generators or doing deep knee bends and connected to devices that generate electricity or something, or running in little squirrel cages in your SUV to make it go. No, they are the sources of energy that we human beings have learned, technologically, how to tap. I like to think of the sources of energy as divided into two classes. I'm going to talk more about these two classes in sort of fundamental resource terms in the next lecture. Some of them are fuels. What's a fuel? A fuel is a substance that stores energy. Gasoline is a fuel; coal is a fuel; uranium is a fuel. These substances contain stored energy, which we have figured out, technologically, how to release and turn into the energy of motion, or the energy of electricity, or the energy of hot water, or whatever other form of energy we want. Those are fuels, substances that store energy.

Others are flows; they're tapped from natural flows of energy. There's a natural flow of energy that comes from the Sun, and it is very substantial. We've already talked about it a lot because it's what powers the climate. We know that it delivers an average of about 240 watts, a rate at which energy is delivered, on every square meter of the Earth's surface. That's one of the natural flows. Other flows of energy are the flow of flowing water, which is carrying with it the energy of its motion, and maybe its gravitational energy if it's flowing from higher to lower. That's another example of a natural flow. That one happens to be driven by the sunlight as part of the hydrological cycle we talked about before. The wind—also ultimately driven by solar heating—is another flow of energy that we can tap. The tides and currents, which I'll talk more about in the next lecture, are also energy sources that we can tap. Those are flows as opposed to fuels. They're natural flows of energy that we can tap into. Fuels are substances we can find, dig up, burn, fission, whatever we do to extract the energy from them. You'll get much more about fuels and flows in the next lecture.

The mix of energy sources differs a little bit from country to country—in fact, substantially in some cases—and I just want to review what some of these energy sources are in the case of the

United States and the world. We're going to look at a graph of that. In this graph that we're going to look at, eventually I'm going to take the United States energy consumption, and I'm going to scale it so that its area on the graph represents the fraction of the energy that the United States uses, which is just about 25% of all the world's energy. Let's just take a look at what some of these energy sources are. It turns out that coal supplies about 23%, almost a quarter, of the United States energy. Oil supplies 39%, almost 40%, of U.S. energy; gas supplies about 23%. Those are the fossil fuels. Add those numbers up, you get 23 and 23 is 46, 85—85% of the energy used in the United States comes from the fossil fuels: coal, oil and natural gas.

Where does the rest of it come from? In the United States, about 8% of it comes from nuclear. That's the second biggest source of energy after the fossil fuels. That's used exclusively to generate electricity. It does propel a few ships, but that's negligible energetically. Nuclear energy supplies about 20% of the electricity in the United States, and about 8% of the total energy in the United States. Hydroelectric power is about 3%. Biomass burning is about 3%; that includes things like wood for heating, wood to generate electricity, burning of waste material to generate energy. And then other is about 1%. If you're a fan of solar energy or wind energy or anything else, all those nice things are crammed into that 1%, and they constitute a rather small fraction of it—most of that other happens to be geothermal energy, which is not a source that we're going to be able to get a lot out of in the future.

That picture is one in which the energy sources for the United States—the energy servants, if you will—are predominantly fossil fuels. Nuclear is a distant second; hydroelectricity is in there, and biomass a little bit. In the world, the picture is just about the same. In the world, coal is at about 24%, oil's at about 39%, gas is at about 24%, nuclear is a slightly smaller fraction of the world's electricity at 6%, and hydro is a little bit bigger at 6%; and the other for the world is also about 1%. Again, when we show those energy sources to scale by area here, the United States energy consumption is just about a quarter, just a hair under a quarter, of the total world energy consumption, for our population of about 300 million people out of a world population of 7 billion, so think about that a little bit.

This pattern of fossil fuels dominating the energy consumption holds pretty much around the world, but there are some exceptions, and it's important to note them because it doesn't have to be this way if a country is endowed with other energy resources or makes other choices. France, for example, gets nearly all of its electricity—about 80% of its electricity—from nuclear energy. They decided to go nuclear as a way of being energy-independent. You can argue whether that's a good choice or a bad choice, but that's a choice France has made. Its pattern doesn't fit these graphs I've shown. Norway, the Congo, and Brazil—rather different countries—all get a large fraction of their energy from hydroelectricity because they have big river systems that are able to supply that energy.

Iceland gets a substantial fraction of its energy from geothermal sources because it's setting on a hot spot where there's a lot of geothermal activity near the surface—volcanic activity, geyser-type activity—and it can get a lot of its energy from that. But if you take the big picture of the world as a whole and kind of forget the distinction between coal, oil and natural gas, which are all carbon-based fossil fuels, the answer is that the world gets about 87% of its energy from fossil fuels. The vast majority of its energy comes from fossil fuels. The energy we use in the United States and the energy that's used in the world; the vast majority is fossil fuel energy.

What's the climate connection? This lecture is entitled "Energy and Climate." What's the connection? The connection is that fossil fuel combustion, and other activities associated with energy, are responsible for 83% of the greenhouse gas emissions that we produce in the United States. There are other sources of greenhouse gas emissions and they comprise about 17%, but energy production produces 83% of the greenhouse gas emissions. The dominant effect, of course, is the production of CO_2 as we burn the fuels, combine them with oxygen, and make CO_2 and water vapor. The CO_2 that's produced in the combustion of the fuels is the dominant effect here, but there are other energy-related effects. Natural gas is methane, CH_4, a strong greenhouse gas. Anywhere in the natural gas distribution system where we have a leak—whether it's near your house or whether it's in a big pipeline—that is an energy-related production of a greenhouse gas. Refineries emit greenhouse gases; that's another energy-related emission. Abandoned coalmines emit methane; active coalmines do too. Those are both examples of energy-related greenhouse emissions and they're included in that 83%.

Some of the sources that are in the 17% that isn't energy-related include things like methane produced from landfills, from sewage treatment plants, and from agriculture. When we produce cement, we work with carbon compounds, and there's some CO_2 released. Cement production is a distant second to fossil fuels in causing atmospheric emissions of CO_2. Cement production and many industrial processes produce various gases, including CO_2, methane and others that are greenhouse gases, so they qualify as non-energy-related greenhouse gas emissions. But again, that's only 17% of the greenhouse emissions in the United States. By the way, these figures also include the effects of so-called carbon sinks; for instance, planting tree plantations and so on to remove CO_2 from the atmosphere. They're already built into this sense that 83% of the greenhouse emissions come from energy production.

What does all this energy buy us? What do we get for using energy at such a prolific rate? Here I'm moving a little bit out of pure science, but I'm going to stay basically quantitative. For many countries, it turns out, there's a pretty direct correlation between the rate at which that country consumes energy per capita—the rate at which its citizens use energy, the amount of energy they use in any given time like a year—and their economic wellbeing. It's measured by the gross domestic product (GDP). You could draw a straight line; you could plot a graph in which per capita energy consumption is on one axis, and per capita gross national product is on another axis, and many countries would lie on a straight line. The United States, Australia, South Korea, Poland, Egypt, Congo all have different amounts of energy consumption, very different, but they have different amounts of GDP as well, and they scale in a proportional fashion, so they all lie on a straight line if you make a graph of per capita energy consumption.

Some other countries lie well below that straight line these are countries—examples of them, there are many—examples of them include Russia and Saudi Arabia. They use a fairly large amount of energy at a large rate per capita, but they don't produce much GNP for it. That means they lie below this average line, and they are, in some sense, energy inefficient. They aren't using their energy very efficiently if you think of energy consumption as something that's helping us to establish a high material standard of living, as measured by a high GDP. There are other countries, though—they tend to be some of the European countries in particular; also Japan—that consume energy at approximately half the rate that the United States does, and yet no one would call those countries places to live that would be sort of difficult because there isn't enough energy. They're not primitive societies in any sense; they're not underdeveloped societies. They simply are using their energy more efficiently. They produce almost as much GDP as the United States per capita, and yet they do so with half the energy.

How are energy and prosperity linked? For a great many countries, they lie pretty much on a straight line. The more energy you consume, the more prosperous you are, as measured by your GDP. Let's take a look at a graph that shows all this together in one place. Here it is, with Saudi Arabia, Russia, Switzerland, Japan, and then a number of countries that lie along a straight line. If I were to produce a more comprehensive graph with more countries, I would still see the same general trend. I would see a scattering of countries, some of which are more efficient—so that they produce more GNP for a given amount of energy—some of which are less efficient. But they would lie on a general trend that suggests the more energy you consume, the more your GDP, the higher your GDP.

Does that mean we all ought to be striving to consume as much energy as possible? I don't think so. I think we need to look carefully at the countries whose GDP is high, but whose energy consumption is low, because those are the countries that have learned how to do things more efficiently. There may be some reasons for the so-called inefficiencies in other countries. For example, the United States is a very large country. It's spread out over a big area, and not surprisingly, the fraction of our energy we spend on transportation is considerably larger than in a small European country, where some other decisions about transportation—like to have very high gasoline taxes, or to have very well-developed railroad systems—may also affect the efficiency with which that country uses energy.

There's also a question of whether the GDP is the right measure of human wellbeing. Here I'm venturing about as far away from science as I'm going to get in this course, but I'm still grounded here in quantitative economics. There are a number of economists who raise the question; maybe the GDP isn't quite the best thing to measure our wellbeing. That's because the GDP includes all the goods and services that we use. For example, if there's an oil spill, and we have to hire people to clean up the oil spill, the employment that we generate hiring the people to clean up the oil spill, that all goes into part of the GDP. Yet we would probably argue that having oil spills is not necessarily a good thing.

Is there some other indicator that provides a sense of human wellbeing that reflects more accurately what things actually go into making our lives better? If there is, how would energy consumption correlate with those things? There are a number of forward-looking environmental economists, particularly, who are trying to formulate better indices of human wellbeing, and to see how different societies stack up against those different measures. One of the measures developed by the United Nations is the so-called Human Development Index (HDI); it's an alternative to the GDP. The HDI has several components. One of them is the GDP, the measure of the total economic goods and services in a country, a measure of its economic wellbeing, its material wellbeing, as measured by money. But there are several other components in the HDI, the Human Development Index. One of them is life expectancy, another is adult literacy, and another one is school enrollment.

What happens if you make the same graph that I showed you earlier of energy consumption rate on the horizontal axis and wellbeing—now as measured by the HDI—on the vertical axis? If you were to make such a graph, you would find a rather different picture. You would find that countries that use very, very little energy per capita have very low HDIs. They are not in very good condition in terms of that measure of human wellbeing. But as you increase the amount of energy consumed, up to a few kilowatts per capita, the HDI—that measure of factors that includes the GDP, life expectancy, adult literacy, and school enrollment—rises rather rapidly. At a few kilowatts, 20 or 30 energy servants per capita, the graph turns over and saturates and levels off, and further increases in the amount of energy per capita consumed don't make any great difference in the HDI. If you believe the HDI measure is a better measure of human wellbeing, then you might say there's kind of a saturation effect. Does energy help us with better lives? Yes, but only up to a point. Beyond that point, increases in energy consumption don't seem to increase our sense of wellbeing, our HDI.

Let me summarize this lecture. First of all, there is a large per capita energy consumption use in the world, about 20 times the human body in North America, over 100 times one human body's worth of energy. Fossil fuels are responsible for the vast majority of that energy, and fossil fuels are also responsible for most of the greenhouse gas emissions. We get something important for that energy; we get a good material standard of living. But it looks like if we use more than 20 or 30 human bodies' worth of energy, we probably don't gain additional wellbeing. That's where we stand. Next, we have to look at what we could do to change that mix of energies away from greenhouse-emitting fossil fuels.

Lecture Eleven
Energy—Resources and Alternatives

Scope:

The fossil fuels that supply most of the world's energy have many deleterious environmental impacts, of which climate-changing greenhouse emissions are but one. Unless we can keep fossil-fuel carbon emissions out of the atmosphere, the challenge of climate change dictates that we find alternatives. Earth's natural systems offer a number of alternatives, broadly classified as *flows* and *fuels*.

Nuclear fission provides nearly greenhouse-free energy, but for the near future, it can't help with transportation. And it faces serious questions of economics, waste disposal, and weapons proliferation. Less developed alternatives have their advocates, but some—such as geothermal and tidal energy—are limited resources. Others, such as nuclear fusion, are too uncertain to count on. Solar-based technologies—including direct use of sunlight and indirect applications, such as wind, hydropower, and biomass—have the advantage of tapping Earth's dominant energy flow. Ultimately, a sustainable civilization will need to use these sources, but they won't be available overnight. For this reason, we need to introduce new energy technologies while using less energy.

Outline

I. If we want to avoid serious climate change and we can't eliminate carbon emission from fossil fuels, then we'll have to find alternative approaches to our energy supply. Earth's natural systems provide us with several distinct energy resources. This lecture is a brief survey of those energy resources and the technologies that exploit them. See Figure 17.

   A. *Flows* are streams of energy that arrive continuously at Earth's surface.

   1. By far, the most dominant flow is solar energy, which constitutes 99.98% of the energy reaching Earth. Solar energy drives subsidiary flows, including wind and flowing water, and through photosynthesis, powers nearly all life on Earth.

2. Nearly all the remaining 0.02% of Earth's natural energy flow is geothermal energy from Earth's hot interior.

3. A tiny fraction of Earth's energy flow is from tidal energy, ultimately originating in the motions of the Earth and Moon.

B. *Fuels* are substances that store energy, generally in Earth's interior.

    1. The *fossil fuels* coal, oil, and natural gas are "stored sunlight," their carbon-based chemical structure storing solar energy that was trapped tens to hundreds of millions of years ago through photosynthesis.

    2. *Nuclear fuels* are substances that can release energy through nuclear reactions. Of the naturally occurring nuclear fuels, we currently use only uranium to produce energy, through the process of nuclear fission. But deuterium, a heavy isotope of hydrogen in ocean waters, carries almost unlimited energy that could be released in the process of nuclear fusion.

II. We have technologies to tap most but not all of Earth's energy flows and fuels. Some technologies are more mature and more economically viable than others. And some of Earth's energy resources are simply inadequate to supply more than a small fraction of humanity's current energy demand.

A. Combustion of fossil fuels is an established technology for both stationary and transportation energy.

    1. Efficiency of large-scale fossil-fueled electric power generation is typically around 40%.

    2. More advanced fossil technologies approach 60% and may cogenerate both heat and electric power.

    3. Fossil-fueled internal combustion engines for vehicles are only about 15% efficient in converting fuel energy to mechanical energy of the wheels.

    4. Fossil fuels, except coal, are in limited supply, and all produce the greenhouse gas carbon dioxide, as well as a host of harmful pollutants.

B. Geothermal and tidal energy flows, although significant in a few localized regions, simply aren't adequate for humanity's energy supply.

C. Nuclear energy currently supplies about 6% of the world's energy, nearly all of it in the form of electricity, making it one of two alternatives to fossil fuels that have demonstrated their ability to produce energy on large scales.

  1. Nuclear energy may be produced either from *fission* of heavy elements, such as uranium, or *fusion* of light elements.

  2. Nuclear reactions are concentrated sources of energy, releasing about 10 million times as much energy as chemical reactions, such as the burning of fossil fuels.

  3. So far, we've learned to harness only nuclear fission for useful energy generation. Fission of uranium powers our nuclear power plants.

  4. Nuclear fission produces radioactive waste, some of which will remain dangerous for tens of thousands of years or more. But the quantities are relatively small compared with the effluents from fossil-fuel combustion.

  5. Although there is considerable anxiety over the safety of nuclear energy, nuclear power plants almost certainly have far fewer detrimental effects on human health and the environment than comparable fossil fuel facilities.

  6. But issues of long-term waste disposal and, especially, the relation of nuclear power to weapons proliferation, cast a cloud over nuclear power's future.

  7. Finally, nuclear fission as it's used today cannot meet the energy needs of our transportation systems.

  8. Energy from nuclear fusion remains an elusive technological goal, although fusion research is slowly advancing. Technologically and economically viable fusion reactors would give us almost unlimited energy with less environmental impact than either fossil fuels or nuclear fission. As a fusion fuel, each gallon of seawater contains the energy equivalent of about 300 gallons of gasoline!

D. Direct use of solar energy taps a flow that carries some 10,000 times more energy than humanity now uses.

  1. Simple *passive solar* technologies allow for heating, cooling, and lighting of buildings using sunlight.

2. More complex *active solar* technologies use solar heating for space heating, water heating, and even the generation of electricity.
3. *Photovoltaic cells* offer direct conversion of solar energy into electricity with no moving parts.
4. Despite their long-term promise, direct solar energy technologies today are generally not economically competitive with conventional energy sources—especially for photovoltaics. But rapid progress may be changing that; worldwide, installed solar generating capacity is growing at some 40% per year.

E. Indirect solar energy sources include water, wind, and biomass.

1. Hydropower taps that portion of the solar flow that drives Earth's water cycle, using the energy of moving or falling water in rivers and at dams. Hydropower is tied with nuclear energy as the second most important source of humanity's energy after fossil fuels and is the other mature energy technology that has been demonstrated to work on a large scale. But most of the world's hydropower has long since been tapped, and new installations often involve major environmental and social disruption.
2. Wind power has considerable potential for generating electricity, and its use is growing rapidly. In the United States, the wind resource is more than enough to supply all the country's electrical energy—but because wind is intermittent, it's not practical to rely on wind for more than 20% of the electricity supply.
3. Combustion of biomass does not make a significant contribution to atmospheric greenhouse gases, provided the biomass is harvested sustainably and without substantial use of fossil fuel. Some biomass systems meet this criterion; others do not. For example, ethanol for motor fuels takes nearly as much fossil energy to produce as is obtained from its combustion. Liquid and gaseous biomass fuels are the one example of indirect solar energy that can be used to power transportation vehicles.

F. What about hydrogen?

1. Although hydrogen is, in many ways, an ideal fuel, producing only water when burned, Earth's energy resources don't include hydrogen.
2. Hydrogen must, therefore, be made from other energy sources, which means that the environmental benefits or disadvantages of hydrogen are largely those of the energy sources used to produce it.
3. Hydrogen-powered vehicles might play a part in a long-term energy future, provided hydrogen could be generated by carbon-free means, such as nuclear fission, nuclear fusion, wind, or solar energy.

III. No energy source is completely benign.

    A. The various alternatives to fossil fuels often involve energy-intensive manufacture or the use of toxic substances (solar photovoltaics), waste and safety issues (nuclear power), aesthetic and wildlife issues (wind), or pollutants and competition for cropland (biomass).

    B. All involve some greenhouse emissions, although much less than with fossil fuels. See Figure 18.

Suggested Reading:

Smil 2006, chapters 5–6.

Going Deeper:

Wolfson, chapters 5–11.

Smil 2003, chapters 4–6.

Web Sites to Visit:

National Renewable Energy Laboratory, http://www.nrel.gov/. At this U.S. Department of Energy site, you can find out about research into alternative energy technologies.

Intelligent Energy Europe, http://ec.europa.eu/energy/index_en.html. This European Commission site features myriad links to programs and information related to the European Union's vigorous pursuit of non-fossil energy.

Questions to Consider:

1. Supporters of a "business-as-usual" energy strategy sometimes

claim that solar energy isn't practical because there isn't enough of it. Is this true? In what sense is solar energy not an immediate solution to our energy problems?

2. What's the difference between nuclear fission and fusion? What would be the fuel source for a practical fusion power plant?

3. Why do non-fossil energy sources entail some greenhouse gas emissions, albeit much less than with fossil fuels?

Lecture Eleven—Transcript
Energy—Resources and Alternatives

Lecture Eleven: "Energy—Resources and Alternatives." We've seen that humankind's energy use is the predominant reason for our influence on Earth's climate through the combustion of fossil fuels and the production of greenhouse gases, particularly CO_2. If we're going to do something about human alteration of climate—if we're going to slow that or stop that process—we need to do something that keeps the CO_2 generated from fossil fuels out of the atmosphere. I'll talk about some alternatives that might still involve fossil fuels in the next lecture, but for now the most obvious thing we could think of doing is replacing fossil energy sources with other sources of energy. This lecture is a survey of some of the other energy sources available to us. I want to take a realistic view of those sources, something I think doesn't always happen. I want to look at particular energy sources and see whether there is really enough there to supply that colossal energy consumption that I described in the previous lecture and also to ask whether we have the technological and economic means to extract energy from some of the other sources that are available to us.

The question is what are some of those other sources? I want to get very fundamental about that, and I want to take a look at Earth's energy resources. Let's look at a diagram that depicts the flows and fuels that make up Earth's energy resources. Once again, flows are streams of energy that come to Earth and flow around through the Earth's system, and we can tap into those streams of energy. Fuels are substances that store energy, materials that have energy locked into their chemical or nuclear structure, and we can extract that energy. Let's look at these fuels and flows. What are Earth's actual energy resources? The diagram I'm showing you has a huge arrow coming in at the upper left. That's the solar flow—the flow of solar energy to planet Earth—and that flow constitutes 99.98% of all the energy arriving at the Earth system. That's a big number.

Solar energy supplies virtually all the energy that comes to planet Earth. When we talk about our energy resources, at least energy flows—these streams of energy that we can tap—the dominant energy flow to planet Earth is solar energy. Remember that when someone tells you there isn't much solar energy, and solar energy isn't a viable alternative. The rate at which solar energy reaches

Earth from the Sun is a little bit more than 10,000 times that 15 trillion-watt energy consumption rate of humankind, of civilized humanity. That's 10,000 times more energy than we human beings use coming to the Earth from the Sun at any given time. The solar flow is huge. It's 10,000 times humankind's energy consumption, and it is the vast majority of the energy that comes to the Earth system; 99.98% of it.

This diagram I'm showing you talks about those flows in terms of percentages of the total amount of energy coming to Earth. The solar flow is 99.98% of the energy coming to the Earth system. We've found already that about 31% of the solar energy is reflected directly off the Earth, off the clouds, off ice caps, off snow, off deserts, and so on. So, 69% of the energy coming to Earth is solar energy that actually gets into the Earth system without being reflected off. What does that energy do? It drives a number of subsidiary flows. For example, 45% of that solar energy that comes into the Earth atmosphere system gets converted into thermal energy, the energy of warm things; mostly the surface of the Earth, but also, to some extent, the atmosphere.

Some of that energy, about 23%, goes into evaporating water from the ocean and other surface waters. That energy gets stored temporarily in the atmosphere in the form of water vapor. I talked about that before, how water in the vaporous state contains this latent heat, this latent energy, associated with the fact that it's been converted from a liquid to a gas, and we could get that energy back if we re-condensed it into a liquid, or nature can get that energy back if nature re-condenses it into a liquid. About 1% of the incoming solar energy ends up driving the winds and the ocean currents—only about 1%. That's still 100 times humanity's energy consumption that goes into driving winds and ocean currents, making the movements of air and water that constitute those currents.

A very tiny fraction of the incident solar energy, about 0.08%—almost a tenth of a percent, almost one part in a thousand, but not quite—is the energy that's captured by living plants, by green plants, through the process of photosynthesis. That energy flows from the plants into dead material. When the plants die, it flows through the animals because all animals are dependent ultimately on plants for the source of their energy. It flows out of the life system through the decay of dead matter and waste matter. A tiny, tiny, tiny fraction of that energy that's been captured by plants, as I described earlier, is stored underground as the fossil fuels. In today's world, there is also a flow of energy out of the fossil fuels; that's the 7 Gt/year of carbon, an equivalent amount of energy you could calculate. It's about 0.007% of the entire Earth energy flow that comes out of the ground right now as fossil fuels are consumed and burned to human uses. There's a very small fraction of energy, about 0.0005% of the total energy flow, that is being extracted from the ground in the form of nuclear fuels and also driving human activities.

There are a couple of other flows that we ought to talk about, too. They're very small compared to the solar flow because the solar flow constitutes 99.98% of all the energy coming to Earth, but they are important. One is the geothermal flow. The geothermal flow is about 0.025% of all the energy coming from Earth. These are round numbers, so they don't add up perfectly, but they're close. The geothermal flow, about 0.025% of all the energy coming to the surface of the Earth, is coming from Earth's interior. This is energy that has two ultimate sources, and geologists aren't absolutely sure how much is one source and how much is the other. The two sources are the primordial energy that existed when the Earth first formed 4.6 billion years ago by the accretion of smaller bodies in the solar system that crashed together under the influence of gravity, heating up and giving the Earth its primordial heat.

Some of that material included radioactive elements like uranium and thorium. Those elements, when they decay, produce heat, and some of that heat is coming from the decay of radioactive elements in the Earth's interior. That's what keeps the Earth's interior warm, and that drives a very, very modest flow of energy to the surface; far smaller than the rate at which solar energy is coming in, a very small fraction. But that energy is significant and has some important effects on the planet. For example, it's ultimately what drives continental drift, which is a major process in reshaping the surface of the Earth and, as we've seen, also may have an influence on climate. It also drives volcanism and processes like that that are also important in climate.

Finally, a very, very tiny fraction of all the energy, 0.0017% of the energy, getting to the surface of planet Earth, is coming from the mechanical energy associated with the rotation of the Earth and the revolution of the Moon around the Earth, from the Earth-Moon system, mechanical energy of the Earth-Moon system. What that mechanical energy does is drive the tides, sloshing the waters of the ocean back and forth, and most places giving us two high tides and two low tides a day. That energy is also available to us to harness, and that energy has a third distinct source, namely the mechanical energy of the Earth-Moon system, but it's a very, very, very tiny fraction of the overall flow.

Let me summarize that in case you came into this course or this lecture thinking, "Gee, there are all kinds of energy sources out there that we human beings can exploit. Let's just try exploiting different ones." The answer is not quite so simple. There are really only three sources of energy to planet Earth—ultimately. One is the solar flow, energy coming from the Sun; one is energy coming from within the Earth—that includes this geothermal heat, and it also includes energy stored as fuels inside the Earth, some of which are fossil fuels, which ultimately came from sunlight, and some of which are nuclear fuels, which were built into the nuclear structure of the materials from which the Earth formed. Finally, we have the tidal flow. That's it. There are really three fundamental sources of energy for us to use. We want to talk about those in realistic terms, about how we might use those to replace the fossil fuels that produce the CO_2 that is largely, but not entirely, responsible for human-caused global warming.

Let's take a look at some of these fuels, or some of these flows, also. We know, in many cases, how to tap into the flows or how to take the fuels and convert that trapped energy into energy we would like to use for transportation, for heating, for industrial processes, for generating electricity, for refrigeration, for cooling, for all the things we do with energy. In some cases, we have the technologies for doing that. In some cases, the technologies are very mature and economical; and in other cases, the technologies are developed, but they aren't developed to become economically viable yet. In other cases, we simply don't yet have the technologies.

Furthermore, some of those sources are more than adequate to supply our energy needs, and some are not. Let's begin by looking in a little more detail at the source of energy we now use to supply most—that is, about 87%—of the world's energy needs. That, of course, is the fossil fuels. We know how to get energy from fossil fuels; we burn them. When we burn them, we combine the carbon in the fuels with oxygen to make CO_2, and we combine the hydrogen in the fuels with oxygen to make water. CO_2 and H_2O are the products of fossil fuel combustion. By the way, how much we get of each depends on the fossil fuel. Natural gas—CH_4 or methane—has got more hydrogens than carbons, so more water forms by burning methane, and less CO_2. Coal, in contrast, is mostly carbon, and so forms mostly CO_2, and oil is somewhere in between. That's why, on a per unit of energy generated basis, natural gas is a better fuel in terms of its climate impact than is coal. Coal is the worst in that term.

Here I want to make a distinction that I think is often muddy. People talk about coal as a dirty fuel. Coal is a dirty fuel.; it produces a lot of other pollutants. Mercury, sulfur, other compounds that come out in the burning of coal are pollutants in the traditional sense of being toxic materials that are harmful to us or the environment. But coal also produces more CO_2 than the other fossil fuels per unit of energy produced, and I don't want to call that dirtiness because, when we burn a fossil fuel, what we want to do is to create CO_2.

CO_2 is not an incidental byproduct that we could remove, that we could avoid—like sulfur is from coal, or mercury. Those are things that happen to be in the coal that we can either remove or trap after burning. The CO_2 is the essential thing we want to make when we burn coal or other fossil fuels. So in that sense, it isn't a pollutant in the usual sense. It isn't a harmful byproduct that doesn't really have to be there, like a toxic contaminant. It's the thing you want to make when you burn a fossil fuel. Keep that distinction in mind when somebody says well, one fuel is dirty and one fuel is clean. Are we talking in a climate sense, or are we talking in a pollution sense? Sometimes those two things go hand in hand, but sometimes they don't.

How do we generate energy from fossil fuels? We burn then and we produce CO_2 and water vapor, as I've said. How well do we do that? Large-scale traditional electric power plants—which are one of the primary uses of fossil fuels, particularly coal, but also increasingly natural gas; not much oil is used in that place—large scale electric power plants, traditional ones are only about 40% efficient. That is, for every unit of energy in the fuel, only about 0.4 units of energy come out as electricity. What happens to the other 60%? It's usually discarded into the environment as waste heat. That can be a problem in itself; that's called thermal pollution. Think about that again. We

©2007 The Teaching Company

have a huge amount of coal we're going to burn that contains a lot of energy, and we throw away about 2/3 of the energy, about 60% or more, of the energy that was in that coal. We literally throw it away. We typically throw it into a river.

If it was going to harm the river too much by raising its temperature, we typically run it through those huge cooling towers—which you may associate with nuclear power, but are just as much needed for coal-burning power plants. We cool down that water before we discharge it to the river. That water contains most of the energy that was in the coal—what a waste. By the way, that waste is not just the result of poor engineering; it's also a result of a fundamental law of physics: the second law of thermodynamics. If you've taken my course *Physics in Your Life*, I have a discussion there on the second law of thermodynamics and how it poses severe limitations on our ability to extract energy by any process that involves heat—combustion of fossil fuels, fissioning of uranium, whatever.

There are more advanced fossil technologies that are called combined-cycle power systems. These combined-cycle power systems use a device like a jet aircraft engine, which in itself is a rather inefficient device for converting heat to mechanical energy. But they take the waste heat that comes out of the basically jet aircraft engine—it's called a gas turbine—and they use that to run a conventional steam turbine. That combination results in efficiencies as high as about 60%. Combined-cycle power plants, they're called. We're still throwing away about 40% of the energy. You can do still better, and that's to do something called cogeneration, in which you have a power plant that generates electricity in the context of a situation—maybe a community that needs heat, maybe an industry that needs steam for some industrial process—and you use what would otherwise be waste heat for these processes that require heat, but not necessarily electricity. That's called cogeneration, and that's an even more efficient thing to do. There are whole cities in Europe that are heated with the waste heat from power plants. There, you're extracting the energy and putting it to good use, and the power plant is therefore much more efficient in using the energy stored in the fuel.

Another major use, of course, of fossil fuels is in powering most of our transportation vehicles. Nearly all our transportation, with the exception of electric trains and a few nuclear-powered ships, is powered by fossil fuels—another exception would be vehicles powered by biomass, for example. But the vast majority of the world's fleet of vehicles are powered by fossil fuels. They're only about 15% efficient at converting the energy stored in the fossil fuels to the mechanical energy of the wheels of the car. This sounds dismal, but it's also a reason for hope because it means we could do a lot better in using fossil fuels efficiently.

Let me mention one other thing about fossil fuels. All the fossil fuels are in limited supply. For coal—maybe not so much for coal—we have enough coal to last a number of centuries. But the remaining reserves of oil—there's a lot of debate about this, and I don't think the debate has been resolved clearly— there are a lot of debate but the remaining reserves of oil and probably of natural gas also, are most likely measured in decades, not centuries. You can quibble about when we will reach the peak of oil, people talk about. But, there will come a time when the oil production in the world peaks and begins an inexorable decline. If the demand for oil is continuing to rise, we will have a real energy crisis then; not when we run out of oil, not when we pump the last drop from the ground, but when that peak occurs.

That peaking occurred for the United States in the early 1970s. Demand continued to rise, and we made up that demand by importing oil from elsewhere. That's why the United States imports such a huge amount of oil every day. If we were the whole world, though, there is no other place to import it from. I don't frankly know whether we're going to have a crisis over the peaking of oil or not. There are some people who are sure that crisis is imminent. There are others who would argue we will slide away from that crisis by switching to alternative sources of energy, both within the fossil realm and outside the fossil realm, and we'll never confront that crisis. Most people would argue we will not confront that crisis for coal because we will never be able to burn all the coal in the ground because of the effect on the climate will be too great. That's fossil fuels. They cause regular pollution, they cause climate change, and they are in limited supply.

I mentioned the geothermal and tidal energy flows. We can, in some limited places around the Earth, extract significant amounts of energy from the ground, particularly in areas that are geothermally active, which unfortunately also tend to be among our scenic areas.

Imagine tapping into Yellowstone National Park, for example. We can supply a significant amount of energy in those localized regions. There are some issues; there are pollution issues, there are groundwater depletion issues, there are land subsidence. These are not ideal energy sources, and they are limited in the regions where we can use them. They cannot ever make a major contribution to the Earth's energy supply.

An exception would be if we learned to dig very, very, very deep into the hot rocks—very, very far down—and extract energy from them. Those hot rocks underlie the entire planet, but they're very hard to get to technologically, and I doubt that that's going to be a viable energy source in any reasonable time. Even then, the overall flow of geothermal energy is still small. It may be larger than what we human beings use, but not much larger, so it's not really going to help us out all that much. Geothermal power, in my mind, although it may be the answer in a few localized places, is simply not the answer or a viable answer.

Nuclear fuels are another story. Today, nuclear energy supplies about 6% of the world's energy and, as we saw, about 8% of the United States' energy. Of the two non-fossil energy sources that are in widespread use, it's one of two where the technology is developed and established, we know how to make it work. We can argue about whether it's economical. We can argue about whether it's environmentally sound, but we do know how to make energy from nuclear fuels, at least in one of two possible ways. If you look at the energy that's stored in the atomic nucleus, you will find that very light nuclei—like hydrogen and helium—don't have a lot of energy stored in them. It didn't take a lot of energy to make those nuclei.

Very heavy nuclei—like uranium—also took less energy to make than the intermediate nuclei like iron, for example. Iron is the most tightly bound nucleus. In the making of iron, a huge amount of energy was released. What this means is if you take a heavy nucleus like uranium and split it in half, you will release energy in producing elements that weigh sort of more middle range like iron. Similarly, if you take lightweight nuclei like hydrogen and fuse them together, you will release a lot of energy as you produce heavier nuclei. That process is called fusion; the former process is called fission. We know technologically how to harness fission to generate energy, particularly to make electricity. That's where that 6% of the energy from nuclear sources is coming from. We do not know yet how to make fusion energy do anything useful for us, except blow things up with thermonuclear weapons. We do know how to make it work there. Nuclear fusion, by the way, is what powers the Sun, and so that's the ultimate source of most of the energy that comes to Earth.

We do know how to make nuclear fission work, and there are some 400 nuclear fission power plants worldwide. They produce highly radioactive waste, and a lot of people are focused on that problem. It is a problem; on the other hand, a large nuclear power plant, in the course of a year, if you compacted its waste, that waste would probably fit under the table I'm speaking in front of. Whereas a coal-burning power plant of comparable size to a large nuclear plant produces 1,000 tons/hour of CO_2—a thousand tons/hour. That's indicative of a difference between nuclear energy and chemical energy, the kind of energy in fossil fuels. It's a difference of a factor of about 10 million in the concentration of energy.

That's both the great thing about nuclear energy and it's also the danger of nuclear energy. This very highly concentrated energy allows a nuclear power plant to be refueled once every year or so with a couple truckloads of uranium; while a fossil-fueled power plant is refueled several times a week with 110-car trainloads of coal that are coming from, in the United States, places like Wyoming where they're being strip mined; or places like West Virginia where the whole tops are being removed from mountains to get at this coal. There's a lot of anxiety over nuclear power. I personally think some of that anxiety is a little bit displaced. But my biggest worry about nuclear power is its connection to nuclear weapons and the possibility that any society or group that has access to nuclear power technology. It isn't the same as nuclear weapons technology—but they have some of the expertise to make weapons.

I personally believe nuclear power is a lot less harmful for the environment than burning fossil fuels, particularly coal to make electricity versus nuclear fission. But all bets on that opinion would be off if we were ever to have a nuclear war that resulted from nuclear power giving people the wherewithal to build nuclear weapons. I think nuclear energy, although it could replace some of our fossil fuels, has some problems we need to think about. In particular, nuclear fission—nuclear power in general—cannot, at least at present, replace our transportation fuels. Maybe in the future, with hydrogen fuel cell cars, we can make hydrogen from nuclear energy, but not yet.

Nuclear fusion is a different story. We don't know how to make nuclear fusion work, as I said, except in thermonuclear weapons. We're working on the problem. We've been working on the problem since the 1950s. We keep saying it's a few decades off, and it keeps being a few decades off. We are now making some real progress. I don't have time here to go into the details of the fusion technology, but let me simply say that if we got nuclear fusion to work, then every gallon of sea water, the special isotope, deuterium of hydrogen in every gallon of sea water would be enough to make a gallon of sea water the energy equivalent of 300 gallons of gasoline. Nuclear fusion, if we ever get it going, has considerable promise. It may end up being too expensive. It does produce some nuclear waste; not nearly as much as our nuclear fission plants produce, and not nearly as long-lived. It's a real possibility, but I think for a very distant future. If we have a climate change problem, we've got to solve that problem long before that distant fusion future.

What about solar energy? I've argued that the solar flow is more than 10,000 times the energy that humanity now uses. We have simple passive solar technologies like houses that have a lot of south-facing glass and not much north-facing glass, and lots of insulation that can get a significant fraction of their residential heating and lighting energy from the Sun. More complex active solar technologies do things like circulate water through simple thermal solar collectors and supply domestic hot water. In cloudy Vermont where I live, I get 95% of my hot water from May to October from a solar domestic hot water system built into the roof of my garage. Another thing you can do with solar energy is concentrate it with a mirror-like device, boil water or some other fluid, and drive a conventional turbine or some other kind of engine, and generate electricity that way.

Perhaps the most exciting and promising solar application is photovoltaic devices. These are semiconductor devices that use silicon technology to convert sunlight energy directly into electricity, with efficiencies that are typically around 15%, but could get as high as about 40%. They can be done as rooftop applications on individual homes or buildings, or in grid-connected power plants up to some tens of megawatts or so; not a huge amount of power, but significant. It's spread out in either sunny regions or not so sunny regions; nevertheless, they will convert sunlight directly into electricity. Solar photovoltaics, and most forms of solar energy, are not, at this point, economically competitive with other energy sources. That's partly because of our tax structure and some of the incentives we give still to fossil fuels, but it's also because we simply haven't developed those technologies enough. Nevertheless, solar photovoltaic installations are currently growing worldwide at about 40% a year.

There are also indirect solar applications. Hydropower is indirect solar energy that we use to evaporate water. Wind, as I mentioned, about 1% of the incoming sunlight goes into driving the winds. When we harness wind energy with wind turbines, we're tapping into that flow. Finally, biomass: the growing of crops or trees, or burning of wood or whatever. Living materials capture sunlight. They're not very efficient at it—they're only 1% or so efficient—but they grow all over the planet, and if we harvest that, we end up being able to burn it. For instance, we have wood-fired electric power plants that have huge piles of wood chips around them. The power plant burns those wood chips and operates like a fossil-fueled plant. In fact, some plants can switch back and forth between the two.

We're working on biofuels to generate gasoline-like substances that we can burn in our cars. Ethanol, particularly, is used in the United States to supplement gasoline. The way we produce it in the United States is not particularly efficient. It takes almost as much energy, sometimes more, to produce it. But Brazil, with its tropical climate, grows a lot of sugar cane, converts it to ethanol, and uses that to power much of its vehicle fleet. Biomass offers considerable promise. Biomass, although it's being burned and producing CO_2, if it's harvested sustainably, and without a lot of fossil fuel involved in processing it, it can become almost carbon-neutral, meaning the trees you replant after you cut them down—or the corn you replant after you cut the corn down or the sugar cane or whatever—then takes as much CO_2 out of the atmosphere as was put in by burning the previous crop. Combustion of biomass does not need to be a net carbon source.

I haven't mentioned one of the favorites of alternative energy people, and that is hydrogen. What about hydrogen? Hydrogen is an ideal fuel in many ways. It burns very cleanly. It produces only water vapor, which is a greenhouse gas, but not to the extent CO_2 is from burning fossil fuels. It doesn't produce other pollutants, but here's the rub. There is no hydrogen energy on Earth. Hydrogen has to be in the form of molecular hydrogen, H_2, and all the hydrogen on Earth is also "burned." It's mostly tied up in water and other compounds that are basically burned. There is no hydrogen fuel. Hydrogen is not an energy source for us in the sense that other fuels are. It's an energy carrier. It's something we could make with other energy sources. We can use other energy sources to split water into hydrogen and oxygen. Then we can recombine them by burning, or by using fuel cells—devices that combine them and convert the energy directly to electricity.

There are many things we could do if we had hydrogen, but we don't have any hydrogen, and we have to make it. The climate impact, and the environmental impact, of hydrogen is the same as the environmental impact of however we make the hydrogen. If we choose to run a coal-fired power plant and use the electricity from that power plant to make hydrogen, then we haven't done any better. Because of inefficiencies, we've probably done a little worse than if we had used the coal-produced energy directly. If we choose to make hydrogen with a nuclear fission reactor, then tied up in that hydrogen are all the environmental problems that are associated with nuclear fission reactions, for example. If we could perfect fusion and use fusion reactors to generate electricity, or use their heat directly to split hydrogen from water, then we might have an interesting source of hydrogen that would be fairly benign.

Or better yet, if we could find a way to take sunlight and directly break up water, which there are ways to do—it's called photolysis; not a very efficient or economical process—then we could make hydrogen directly from sunlight energy, or we could built photovoltaic power systems, use the electricity to split water into hydrogen and oxygen, and we'd have hydrogen. Hydrogen would be a great energy carrier in some future economy. It's going to be a long time coming. It's not going to come soon enough to solve our greenhouse warming problem, and we have to remember there isn't any hydrogen fuel available to us. We have to make it, and we have to make it in a benign way.

Finally, let me end by reminding you of something else about all these energy sources. Although many of them are better in terms of pollution and climate impact than the fossil fuels, no energy source is perfectly benign. I want to show you, at the end of this lecture, a graph that looks at greenhouse emissions from different energy alternatives and scales them relative to each other. For the fossil fuels, we're pretty sure we have a good handle on that. I'm showing you, for the fossil fuels, low and high values of actual technologies in terms of the amount of CO_2 emitted in grams/kilowatt hour of energy coming out of these fossil fuels. The worst is coal; the next is oil; the next is natural gas. Those are the fossil fuels. We expect that the fossil fuels are bad in terms of producing CO_2, if I can call that a bad thing. I think I can because we're worried about global climate change caused by that CO_2.

Next come solar photovoltaics. They're much harder to estimate, but the energy used in making the silicon, the high purity silicon that's used in making semiconductors, consumes a lot of fossil fuel. The high estimate for photovoltaic energy is that it is a significant fraction of the CO_2 produced from natural gas because of the fossil fuels used in its manufacture. Most of us would argue that it's actually a lot lower, and the low estimate is very small. As we develop new photovoltaic technologies that number will go down. Hydroelectricity, in the temperate zone, can be very, very free of greenhouse emissions. But in the tropics, it can produce large amounts of greenhouse emissions, as I mentioned, from the methane that's being burned. Biomass can be done either sustainably or not so sustainably, processed with a lot of fossil fuels or not, and may or may not be greenhouse neutral and nuclear may produce some greenhouse emissions, largely from the processing of nuclear fuels and the building of concrete to make reactors. No energy source is benign. But all the energy sources, other than fossil fuels, are a lot better, at least from a climate standpoint.

Lecture Twelve
Sustainable Futures?

Scope:

Earth's future climate depends not only on science but also on human behavior. Climate models run with different scenarios for human population and energy use yield very different projections of future climate. Avoiding disruptive climate change probably means keeping atmospheric carbon dioxide to at most a doubling of its pre-industrial level. We can achieve that goal only through a substantial reduction in anthropogenic carbon emissions. Here, we have choices: We can continue to burn fossil fuels but capture and sequester the carbon dioxide produced. We can switch to energy sources that produce less carbon dioxide. We can use less energy. Practical strategies for mitigating climate change will most likely involve all three approaches. There are many paths to a stable climate, and we have the luxury of choice. But we don't have the luxury of time.

Outline

I. With regard to climate change, humanity's goal is "stabilization of greenhouse gas concentrations in the atmosphere at a level that would prevent dangerous anthropogenic interference with the climate system."

   A. This goal was adopted at the 1992 Earth Summit in Rio de Janeiro and signed by nearly 200 countries, including the United States. The goal is embodied in the United Nations Framework Convention on Climate Change (UNFCC).

   B. Most climate scientists suggest that a threshold for "dangerous anthropogenic interference with the climate system" is somewhere around a doubling of pre-industrial atmospheric carbon dioxide.

      1. The pre-industrial concentration was about 280 parts per million; therefore, a doubling would be 560 parts per million.

      2. Today, we're at around 390 ppm, climbing by about 2 ppm per year.

3. A look at the IPCC scenarios suggests that we won't avoid "dangerous anthropogenic interference" if we continue with "business as usual." See Figure 19.

II. Given that the dominant anthropogenic effect on climate is from combustion of fossil fuels, we have four different ways to avoid that "dangerous anthropogenic interference."

  A. With global engineering projects, we could attempt to counter the effects of increasing greenhouse gases. But we really don't know enough about the complex workings of all Earth's systems to make this approach a wise one. Approaches that have been discussed include the following:

  1. Injecting sulfur particles into the upper atmosphere to reflect sunlight.

  2. Placing mirrors in orbit to reflect sunlight.

  3. "Seeding" the oceans with iron to increase phytoplankton growth and, thus, take up carbon dioxide.

  B. We could continue to burn fossil fuels but prevent the resulting CO_2 from entering the atmosphere. This approach entails some form of *carbon capture and sequestration* (CCS).

  1. CCS with conventional power systems is technologically and economically challenging because of the huge volumes of carbon dioxide produced—about 1000 tons each hour from a typical electric power plant. Furthermore, the carbon must be separated from large volumes of nitrogen that constitute most of ordinary air.

  2. Newer schemes envision removing carbon before fuel is burned, typically by gasifying coal to form a mixture of carbon dioxide and hydrogen. The hydrogen would be burned, and the CO_2 would be sequestered.

  3. Sequestering carbon means storing it where it can't escape to the atmosphere. Brine-containing sedimentary rock formations, abandoned oil fields, and the deep ocean have all been suggested as sequestration sites.

  4. CCS can work only with large, stationary sources of CO_2, such as electric power plants and industrial boilers.

  5. CCS has both economic and energy costs. It would take as much as 40% more fuel to produce a given amount of energy with CCS as opposed to conventional technology.

C. We could switch from fossil fuels to the various energy alternatives discussed in the preceding lecture.

1. Of those, only nuclear fission and hydroelectricity have demonstrated potential as large-scale energy sources, and both are currently suited only to electric power generation. The hydro resource is almost fully exploited, and an expansion of nuclear power faces waste disposal, safety, and weapons proliferation issues.

2. Other alternatives, especially direct and indirect solar technologies, have considerable promise—but most aren't technologically or economically competitive with cheap fossil fuels at present.

3. A distant future might feature a carbon-free energy economy using solar energy or nuclear fusion, with hydrogen produced for transportation fuel. But that's not a near-term solution.

D. We could use less energy.

1. This doesn't mean "freezing in the dark" but, rather, using energy more intelligently—getting energy's benefits without using as much energy.

2. In the United States, we've already reduced our *energy intensity*—the energy required per unit of gross domestic product—by some 50% since 1975.

3. During energy shortages, we've demonstrated our ability to reduce energy consumption. Had we continued such reductions, we would today be using far less energy and producing far lower greenhouse emissions.

4. Many of our energy systems can be made far more energy efficient.

5. Efficiency gains compound. For example, better home insulation means downsizing the heating system, lowering costs and energy required to manufacture the system. "Superinsulating" may eliminate the need for a heating system altogether. Similar gains abound in industrial applications.

III. There's no one solution to anthropogenic climate change. Instead, we'll need to use a range of energy options to reduce our greenhouse emissions.

  A. Here, we have choices—more than we actually need!

  B. A study by Stephen Pacala and Robert Socolow of Princeton University shows how the use of "wedges" from a wide range of choices can stabilize our greenhouse emissions. See Figure 20.

  1. In 50 years, each wedge will have prevented 7 gigatonnes of carbon from entering the atmosphere.

  2. Wedge choices include improving energy efficiency, substituting nuclear for coal-fired electric power plants, substituting natural gas for coal, using biofuels for transportation, implementing carbon capture and sequestration, planting trees in temperate climates, building 2 million wind turbines to replace coal-fired electric power plants, providing 2000 gigawatts of solar photovoltaic power production to replace coal, halting tropical forest deforestation, and restoring 300 million acres of damaged forest.

  3. The wedge approach isn't the complete solution to anthropogenic climate change, and it alone won't stabilize atmospheric CO_2. What the wedge approach does is to stabilize our emissions over the next 50 years. After that we need to start reducing emissions if we're to stabilize CO_2 levels at less than a doubling of its pre-industrial level.

IV. Review of the entire course.

  A. Global temperature is rising.

  1. The increase of recent decades is almost certainly attributable to human activities, especially the emission of carbon dioxide and other greenhouse gases. CO_2 levels are now some 30% higher than anything the planet has seen in hundreds of thousands to perhaps millions of years.

  2. Other climatic changes also point to an anthropogenic origin for present-day climate change.

  B. Computer models suggest an additional rise of 1.5°C to 4.5°C over the coming century. Impacts include rising sea level, more extreme weather events, droughts and floods, changes in species ranges, spread of tropical diseases, and less likely, sudden "surprise" changes in climate.

C. Most of humanity's impact on climate comes from our combustion of fossil fuels for energy, which necessarily produces the greenhouse gas carbon dioxide.

    1. We will need to curtail atmospheric greenhouse emissions if we're to avoid "dangerous anthropogenic interference with the climate system."

    2. Many approaches can work, including a mix of alternative energy sources, greater energy efficiency, and carbon sequestration.

    3. But we need to act now.

Suggested Reading:

Houghton, chapter 11.

Wolfson, chapter 16.

Going Deeper:

Pacala and Socolow.

IPCC 2005.

Lovins, et al.

Web Sites to Visit:

Princeton Environmental Institute Carbon Mitigation Initiative, http://www.princeton.edu/~cmi/resources/stabwedge.htm. This Princeton University site provides a detailed description of the "wedge" approach to greenhouse stabilization, including a movie, a simulation game, and links to the original papers.

Questions to Consider:

1. "Using less energy" often conjures up images of people shivering in bulky sweaters or crammed into impossibly tiny cars. Why are such images misleading?

2. Which seven wedges would you choose to stabilize our greenhouse emissions?

Lecture Twelve—Transcript
Sustainable Futures?

Welcome to Lecture Twelve, the final lecture of this course, entitled "Sustainable Futures?" with a question mark after that title. Why the question mark? Because I want to ask the question of how might we build a sustainable future, a future in which we have a stable and benign climate, one that doesn't cause all kinds of disruptions and dislocations for the various species that live on Earth, including particularly our own. What's our goal here? We're faced with the problem of anthropogenic climate change; climate change caused by human activities; predominantly the burning of fossil fuels for energy, but also other activities as well. Humankind as a body has, in fact, articulated a particular goal, and that goal was articulated at the 1992 Earth Summit in Rio de Janeiro, which started the process called the United Nations Framework Convention on Climate Change (UNFCCC). The Kyoto Protocol, which you've surely heard about, is a sub-aspect of the UNFCCC.

What is the goal? The goal, as quoted in the UNFCCC—as it's called—is, with regard to climate change, our goal is "stabilization of greenhouse gas concentrations in the atmosphere at a level that would prevent dangerous anthropogenic influence with the climate system." We're trying to prevent dangerous anthropogenic influence with the climate system. That's all well and good, but what is dangerous anthropogenic influence? That's a matter of some debate, but not a whole lot. Most client scientists would suggest that somewhere around a doubling of pre-industrial CO_2, we reach a level where the interference that humankind has made with the climate system—in other words, the climate change that will have occurred as a result of a doubling of pre-industrial CO_2—that will be dangerous in the sense that it will be disruptive to the systems, like agriculture, that keep us going, to the marine ecosystem, to the climate that keeps us healthy, and so on.

Somewhere around a doubling of pre-industrial CO_2 concentration will result not just in unpleasant climate change—some places it will be pleasanter, but some places it will be unpleasant—but will result in real dangers to human societies, and to other species as well. I want to pause a minute and say something right at the outset. I personally don't think global climate change will ever be the kind of disaster that, for example, an all-out nuclear war would be, or a strike from a large asteroid, or even a moderate-sized asteroid. These are things that are potentially civilization-destroying events. I don't think global warming fits in that category. There are some people who would argue it does, and who are very alarmist about it, and think this is an enormous crisis. I think it's a big crisis possibly, but I don't think it's a civilization-destroying event.

For one thing, it's happening rather slowly. Unless one of those surprise events occurs, it's something that we *can* adapt to. It may not be very pleasant to have to adapt to it, but it's something the human species and human civilization can survive. That doesn't mean I think it should be allowed to happen, but I don't rank it in the same category as an all-out nuclear war or an asteroid impact. Most scientists would argue that somewhere around, again, a doubling of pre-industrial CO_2 is where we really get dangerous with climate change, with anthropogenic influence on climate. Some would argue that the threshold is probably a little bit lower. Some might argue it's a little bit higher, but if anything, some would argue it's lower.

Remember what the pre-industrial concentration of CO_2 was. Remember my milk bottles. The pre-industrial concentration was about 280 ppm. Today it's around 390 ppm. A doubling is twice 280, or 560 ppm. We're partway there, but we're not anywhere near all the way there. On the other hand, CO_2 concentrations are going up by about 2 ppm every year, and that rate is almost certain to accelerate as developing countries, particularly China and India— rapidly industrialized China particularly has huge resources of coal, and that's the quickest way for them to get up to our kind of standard of living, by burning a lot of that coal, particularly to make electricity.

We've already looked at some graphs that show the CO_2 emissions and the resulting temperature changes, under several of the IPCC scenarios for what humankind might do. For example, we saw an enormous variation in the CO_2 emissions between the A1FI—the fossil-intensive business as usual scenario—and the more environmentally friendly scenarios, the B categories; big differences in the amount of emissions. We saw rather less difference in the amount of temperature change because the temperature change depends not on the instantaneous rate of CO_2 emission, but on the accumulated CO_2. What if you look instead at not the CO_2 emissions or the temperature change, but rather what the CO_2 concentrations might be in each of those scenarios?

If you were to look at what the CO_2 concentrations predicted by the year 2100 are in those scenarios, you would find that, for example, the A1FI, the fossil-intensive business as usual scenario, predicts the highest level of CO_2 concentration at the end of the 21st century, almost 1,000 ppm. That's in the range of what we think was the CO_2 concentration during the time of the dinosaurs, when much of the Earth was tropical in climate. That's at the end of the present century, in the A1FI, the fossil-intensive, scenario. That's clearly well above, by almost a factor of two, the doubling of pre-industrial CO_2, to about 560 ppm that most climate scientists would regard as dangerous anthropogenic influence on the climate.

Many of the other scenarios, the A2 scenario, for example—the scenario that involved a non-globalizing world, but still with emphasis on economic development; a world in which the population continues to grow until the year 2100—that one doesn't produce quite as much CO_2 concentration as the fossil-intensive business as usual scenario, but it gets us up to about 800 ppm. Several of the other scenarios also get us up above 600 ppm. In fact, only the most benign scenarios, the B1 scenario is the most benign. The B1 scenario is the scenario that says we're going to be a globally-oriented world, but we're going to put our emphasis on environmental protection. That scenario is the best, and it gets us to about 500 ppm, give or take a little bit. It gets us a little bit below that threshold, but just barely, for dangerous anthropogenic influence. Who knows whether that's a sharp threshold or a gradual thing. Even that may be too high. The next best one is the A1T scenario, the economic development-oriented world that is globalizing but uses advanced technology to replace fossil fuel combustion. That one gets us to somewhere between 500 and 600 ppm, just about in that pre-industrial doubling range.

Of all those IPCC scenarios, those storylines, those projections for how humankind might behave in the future, only a couple of them—maybe three—get us in the range of that 560 ppm doubling at the end of the 21st century; and for one of those three, the concentration is still continuing to climb at an increasing rate. For the other two—the A1T scenario and the B1 scenario, the environmentally friendly scenario—they are both leveling off by about 2100, at levels approximately twice the pre-industrial concentration. Those look like viable scenarios for meeting our goal if our goal is indeed to avoid "dangerous anthropogenic influence" with the climate system.

How do we get there? We're not headed there now; I want to emphasize that. We are headed toward the combustion of more and more fossil fuels. Over the next couple decades, it looks like a lot more big power plants, coal-burning power plants, are going to come on line. It looks like a lot more people are going to have cars run on fossil petroleum. It doesn't look very good for getting off that A1FI scenario, yet that's what we've got to do. Given that the dominant anthropogenic influence on climate, the thing we're trying to avoid—a dangerous anthropogenic influence on climate—is from the combustion of fossil fuels, we really have four different ways, as I see it, to avoid that influence. Some of them I mentioned briefly— I'm going to go into more detail on them—and some of them I haven't even talked about yet.

Here's one. I haven't mentioned this at all, and I'm going to be pretty dismissive of it. We could devise global engineering projects that compensate for the upsetting of the Earth's energy balance by fossil fuels—particularly fossil greenhouse gases, CO_2, and methane—by upsetting the energy balance in the other way to compensate for that. We could keep burning fossil fuels as much as we wanted, but we could do things like, for example, by injecting sulfur particles into the upper atmosphere. This has actually been suggested; seed the upper atmosphere with sulfate particles. Serious papers in the scientific literature have been written about this. Put just the right amount to reflect enough sunlight to compensate for the enhanced greenhouse effect due to fossil fuels.

We could place mirrors in orbit to reflect incoming sunlight and reflect just enough incoming sunlight to compensate for the increased warming, again associated with the greenhouse gases.

We could seed the oceans with iron. Most of the open ocean is a vast desert; there isn't much life there. One of the reasons there isn't much life there is because there's a dearth of nutrients; in particular, iron. If we seeded the oceans with iron, we could cause the growth of phytoplankton—tiny marine plane organisms that do photosynthesis—and they would take up CO_2 out of the atmosphere and we could remove some of the CO_2 that we put in.

Those are three of a number of examples that have been proposed of what I call global engineering projects. We're trying to reengineer the whole Earth system. I personally, as a scientist, don't think they're very good ideas. I don't think they're very good ideas because the Earth system is very complex and we don't fully understand it yet. We are doing major tinkering with it, with any of these approaches, and we don't know if they're going to backfire; if they're going to trigger positive feedback mechanisms that make things worse. We don't know enough to know how those global engineering projects would really work. I think it's dangerous to play with planet Earth in that way, and so I'm going to dismiss that as one of the options. Proponents of these options would argue differently, but I don't think global engineering is the way to solve the problem, but it would allow us to keep burning fossil fuels.

Now, here's another way we could continue to burn fossil fuels, which we seem to be good at doing, we know how to do, and we have the technology for doing. We could burn the fossil fuels, but we could prevent the CO_2 from getting into the atmosphere. This approach entails something called carbon capture and sequestration, CCS. In these CCS schemes, one captures the carbon, either before or after the fuel is burned, and one puts the carbon somewhere— typically deep underground, possibly deep in the ocean—and keeps it out of the atmosphere. How would these work? First of all, I want to remind you that CCS is a daunting problem because unlike the pollutants—sulfur, mercury, other particulates—that we can remove from the exhaust streams of combustion processes in the tailpipes of our cars and the smokestacks of our power plants, CO_2 is what we want to produce when we're burning fossil fuels. There's a vast amount of it being produced.

As I mentioned before, from a typical large power plant—a one-gigawatt power plant, a 1,000-megawatt power plant typical of a medium- to large-size power plant these days—we produce about 1,000 tons of CO_2 every hour. We would have to capture that CO_2 and sequester it. First of all, we have to capture it from an even vaster gas screen because most of the atmosphere is nitrogen, and most of that nitrogen just goes right through the power plant and comes out the smokestack unaltered. A little bit of it combines in the high temperatures to make the nitrogen oxide pollutants—that's a problem too—but most of the nitrogen comes through. So, we have this vast amount of gas coming out of the smokestacks. We have to

©2007 The Teaching Company

remove the CO_2, which is a vast amount itself, 1,000 tons/hour. Then we have to pump it deep underground and somehow keep it out of the atmosphere for a very, very long time. That's a big technological challenge.

People are working on that challenge. One way to make it a little less challenging is to process fuels before they're burned and try to remove the carbon then. Since these fuels contain hydrocarbons, you can then be left over with hydrogen, which you can burn as a fuel, but you've taken the carbon out beforehand, and that's an easier thing to do than remove it from the gas stream after combustion. On the other hand, you've removed some of the carbon, which contains some of the energy. If you combined it with oxygen to make CO_2, you'd get more energy out and you've lost that energy by taking the carbon out that way.

Sequestering means you have to store the carbon where it simply can't escape. Where is that? Scientists believe that one place is in rocks that are saturated with a brine, a kind of saltwater solution. You can pump the CO_2 down there and, with the right geological formations, geologists believe it can stay trapped there for thousands of years without leakage. Obviously, if the carbon leaks back up and into the atmosphere, you haven't done anything, and you've wasted a lot of money, and energy, and other things to do that. But if you could find a geologic formation where it could be sequestered, that would help a lot.

Another place to sequester it is in oil fields. The Earth is full of old oil wells, abandoned oil wells, where we pumped all the oil out. In fact, CO_2 is a commercial commodity for the oil industry. They buy it, and they pump it down into oil wells, where it helps bring up the last residues of oil down in the wells. There's a kind of nice symbiosis there. If you produce CO_2 from fossil fuel combustion, and you capture it, and you pump it down abandoned oil wells, you may bring up more oil and get some more energy out. That's a possibility. On the other hand, regions full of oil fields may have uncharted oil wells that act as possible leaks for the CO_2 to get out into the atmosphere. So, that may not be such a good idea.

There are also those who advocate putting CO_2 deep into the bottom of the ocean where, under this enormous pressure, it would presumably stay. That would solve several problems. It would get it out of the atmosphere. It would keep it out of the surface waters where it's causing acidification problems, as I pointed out. But we don't know enough about the ecology of the deep ocean, or the movement of the CO_2 in the deep ocean, to be sure that that method would work. There are some trial plants that are doing that off the coast of Norway and some other places. There's a trial operation in Canada, taking CO_2 from power plants in the Dakotas, but these are just small pilot projects going on right now. I should mention CCS is obviously a technologically difficult thing to do. It's obviously an economically expensive thing to do, and it also consumes a huge amount of energy. To pressurize CO_2 and pump it deep into the ground or deep under the ocean takes a lot of energy.

A CCS-equipped power plant may require something like 40% more fuel—40% more coal, 40% more natural gas, or whatever—to produce a given amount of electrical energy, say, than it would if it didn't have the CCS technology. But CCS is something that people are looking actively at. I must say, after a report on this by the IPCC—I used to be a skeptic about this. I thought it was just a way of keeping the fossil industry going—now I think it's something we ought to look at as one possible step of many in the solution to the problem. But I don't have high hopes for CCS, and I think we need to wean ourselves from fossil fuels for a number of other reasons, including their pollution effects and their scarcity. I don't think CCS is a long-term solution, but it might be a short-term solution until some of those longer ones can come on line.

A third possibility is that we could switch from the fossil fuels to the various energy alternatives that I've described in the previous lecture. I want to be realistic here. Of those, only nuclear fission and hydroelectric power are currently developed at the large scales that we need to supply a significant fraction of the world's energy. Both are at present, suited only to one thing, namely electric power generation. So of all the things we do with energy—electricity, by the way, is one of the most versatile forms of energy—nuclear fission and hydropower basically are good for producing electricity and not much else. The hydro resource is almost fully exploited in the industrialized countries. You can talk about building small-scale hydro in your state in the United States, and maybe there are places you can do it, but I guarantee you you aren't going to get a great deal more power out of hydroelectricity because almost all the hydro sources are developed. In fact, in some states, we're taking down old

©2007 The Teaching Company

hydrodams because they have ecological consequences we don't like.

The nuclear option is a different story. This is not a course in nuclear power—I could go into the whole thing. But you know there are some problems with nuclear power. I would argue that those problems may not be as serious, with the exception of proliferation, as many people think. But still, there are questions to be raised about nuclear power. There are other alternatives. In particular, there are the direct and indirect solar technologies that I discussed; the conversion of sunlight into electricity through photovoltaics, the conversion of sunlight directly into heat through simple thermal solar collectors, the use of solar concentrators to make heat engines, basically, thermal power plants that run on sunlight. These are all possibilities.

The use of indirect solar energy, wind power, also growing very rapidly at about 30% a year, especially in Europe—Denmark gets about 20% of its electricity from the wind, for example—these are coming along. They have considerable promise. They are advancing technologically very rapidly. Wind power, in some areas, is becoming competitive with ordinary fossil fuel-based grid electric power. I think that's a good thing, but a single wind turbine of the largest size produced today makes perhaps five megawatts of electricity at its peak. A single large coal-burning power plant produces 1,000 megawatts, 200 times as much. You need a lot of wind turbines to replace coal-fired power plants; and the coal-fired power plants run round the clock, and the wind doesn't necessarily blow round the clock. You've got to be realistic about these things. Nevertheless, the direct and indirect flows of solar energy—direct use for photovoltaics and so forth, indirect use of wind, water, and biomass—do have serious potential for replacing fossil fuels. But the most promising of those, at least in the long term, like photovoltaics, are simply not developed, technologically or economically, to the point where we can switch them in.

Having said that—this is as close as I'm going to get to policy in the whole course—we have been subsidizing the fossil fuel industry in many ways for a long time. For instance, we subsidize highways as a means of transportation in the United States over, say, rail, which is also fossil-powered; although if it's electric, it could be nuclear powered. We give oil companies depletion allowances as they remove oil from the ground. There are many things we do that encourage the use of fossil fuels through policy. One of the discussions that's been going around in the IPCC circles, and environmental circles in general, is the idea of a carbon tax. Tax fuels on the basis of how much carbon they put into the atmosphere.

We could do that, and that would be a policy change. A policy change like that would change the economics of where, say, photovoltaics set relative to coal power. It wouldn't be enough, at the present, to reverse that economics, but it would help. My point is there are policy decisions we could make that would take the economic ordering of these technologies and change it a little bit in favor of the non-carbon technologies. You may or may not want to do that politically, and that's not what this course is about, but there are ways to nudge things in more favor of these non-carbon technologies. It isn't as though, right now, it's a level playing field. It's quite different from that if you look at the subsidies, direct or indirect, that we have for carbon-based fossil energy.

I could imagine a distant future in which we have carbon-free energy that's coming from, say, solar photovoltaics or from nuclear fusion, and that produces our electricity. We take that electricity, and we use it to split, as I said before, water into hydrogen and oxygen, and we use that hydrogen as the fuel for our portable transportation vehicles. That's a completely carbon-free energy future, and it could fusion-powered, or it could be solar-powered. I think we could have that in 50 years, but we need to do something well before 50 years to get our carbon emissions under control. That's a goal to have for the distant future, but to be realistic, that's not where we're going to be now. There are a few other things we could do. For example, just shifting from coal to natural gas buys us a factor of two reduction in carbon emissions because natural gas is CH_4, methane, and it's got a lot more hydrogen, so a lot more of the byproduct is water than CO_2. Coal is almost all carbon; it's almost all CO_2 it produces.

There are things we could do in the interim. Natural gas may help us slide over to a completely carbon-free future. We can burn the fossil fuels in more efficient engines, and so on. There's one last thing we could do: We could use less energy. If you're somebody who remembers the late 1970s and 1980s, you can remember all this talk of energy conservation from the energy crises of the 1970s and early 1980s. People shivered in the dark. They put on lots of sweaters;

212

they kept their heat at 55° and all that. That is not what I'm talking about. I'm not talking about giving up the benefits of energy. I'm talking about using energy more intelligently; getting those same benefits from energy, but using less energy. If you don't think that's possible, take a look again at Europe and North America. Europe uses energy per capita at roughly half the rate that North America does.

There are some reasons why there is a legitimate difference. I mentioned, for example, the large size of the United States and, for that matter, Canada, that requires more transportation energy. But there are plenty of other reasons. For example, the high gasoline taxes in Europe that encourage smaller cars, the development of localized village centers rather than suburban sprawl. You name it; there are lots of things going on in Europe, that aren't going on in the United States, that allow the Europeans to use half as much energy per capita without sacrificing their quality of life. That's what I'm talking about, using energy more intelligently.

I don't want to knock North America. We've already done a lot of things to reduce what's called our energy intensity; the energy it's required per unit of GDP, if we want to use that as a measure of our wellbeing, by some 50% since 1975. We have increased our total energy consumption, and we've slightly increased our energy consumption per capita, but we would have increased those things a lot more if we hadn't reduced our energy intensity, the amount of energy it takes to produce a unit of GDP. We've had some energy shortages, in the 1970s and early 1980s. During those shortages, we showed remarkable declines in our oil consumption in particular; our energy consumption in general, but particularly in oil consumption. If you look at the period of several years around 1980, late 1970s to early 1980s, of U.S. oil consumption, you would see a very steep decline, an exponential decline, in U.S. oil consumption. That decline stopped in the early 1980s when the price of oil fell and we no longer felt it was necessary to conserve it; and the oil consumption started right back up again.

But if you continue to trace the exponential fall-off curve downward to the early 21st century, we could be using a tiny fraction of the amount of oil we are now. We wouldn't have to import any oil if we had simply continued to pursue the energy conservation, energy efficiency, measures we were pursuing then. I'm not saying, and I'm not promising, that it would have been technologically or economically feasible to continue all that way down. But it's obvious to me, and to anybody who looks at these numbers, that we certainly could have continued substantially further down, and we could have done that even though the price of oil went down. But the price of oil going down took away the incentive to do so.

Finally, most of our energy systems can be made a lot more efficient. Just to give you some examples, let me consider vehicles. The average vehicle today doesn't get very good mileage compared to the best vehicle you can buy. You can go out and buy a hybrid vehicle whose fuel efficiency is maybe three times that of the typical vehicle in the American car fleet, for example. Car manufacturers have produced prototype vehicles—not things that they could sell economically or that you'd necessarily be comfortable driving—that get up to about 300 miles/gallon of gasoline. There are great possibilities for gains in energy efficiency.

Look at another area of technology. In the 1970s, refrigerators were horribly inefficient. When we began imposing standards on refrigerator efficiency, the typical model in the early 21st century was about three times more efficient. It used about a third the amount of electrical energy as did the typical 1970s model. The best possible model you can buy is about twice as efficient as that. If we move to the most efficient possible devices that have already been proven to be possible, we could reduce our energy consumption enormously. Let me point out that energy efficiencies compound. If you make a house better insulated, you need a smaller furnace; it took less energy to build the furnace, and so on. It's the same in industry. You could really compound these energy savings and do very well in improving energy efficiency.

There is no one solution to the climate change problem. We're going to have to use a range of these options to reduce our greenhouse emissions, and we actually have quite a bit of choice. I want to end by describing a study by Robert Socolow and Stephen Pacala of Princeton University, who have this wedge idea. They say let's try, over the next 50 years, to avoid a certain amount of carbon emissions. I'm going to show you a graph that gives their wedge approach. What we're seeing here is a range of times over about 100 years. The first 50 years of it, we show the historical emissions of CO_2 continuing to rise. Socolow and Pacala project what would happen if they just continued to rise at the same rate for the next 50 years. There would be additional CO_2 emitted into the atmosphere and they say let's prevent that CO_2 from being emitted. That forms a triangular region on this graph of emissions versus time.

They say how could we get rid of that triangular region? How could we keep that CO_2, about 50 Gt, from being emitted? The answer is their wedge approach. Let's divide the triangle up into seven. So, they prevent a total of seven Gt from going in in that final year, and they reduce the total atmospheric emissions by about 50 Gt, our atmospheric content of carbon by about 50 Gt. What are those wedges? That's the beauty of this approach. The wedges are choices. The wedges give us choices; they give us choices. For instance, we could substitute nuclear for coal-fired power plants. That would be a wedge.

If you don't like nuclear power, don't choose that wedge. We could substitute natural gas for coal; that gives us another wedge. We could use bio-fuels for transportation; that gives us another wedge. We could implement CCS on our coal-fired power plants; that gives us another wedge. We could plant trees in temperature climates to pull carbon out of the atmosphere; that's another wedge. We could build two million wind turbines to displace coal-fired power plants. We could build 2,000 gigawatts worth of solar photovoltaic power generation capability. We could halt tropical forest deforestation. We could restore 300 million acres of damaged tropical forest. Any one of those would be one of these wedges, and we need seven of them. We've got a lot more than seven.

If you buy this wedge approach, 50 years from now, we will have stabilized carbon emissions, but we'll still be emitting CO_2. At that point, we have to go downward. This is not the complete solution, but it's an optimistic way of looking at the next 50 years and saying hey, if we take concerted action now, we can really do something about this problem. But the key is now; we have to start now to make this work. If we don't start soon, we're going to end up seeing a tripling of pre-industrial CO_2.

Let me end with a quick look at the entire course. This course is about the rise in the Earth's global temperature, the change in climate. There's been a warming. We've seen it; it's been especially strong in recent decades. Most of it is attributable to human activities, primarily greenhouse emissions from combustion of fossil fuels. It's going to have a significant impact on the environment, on human society. Most of that impact is associated with our consumption of energy and we have many possible approaches for minimizing anthropogenic climate change, but we've got to act soon.

Let me end, characteristic of the entire course, with one more graph. We're going to look at a graph, which is basically that set of 10 reconstructions of the past millennium's climate that showed a gradual decline over the first roughly 900 years, and then the sharp upturn. Imposed on that graph also is the instrumental temperature record, an eleventh curve on that graph, from the last 150 years, showing that sharp upturn. Now we're going to impose on that graph also a 3°C rise by the year 2100—that's about average for the IPCC projections. That graph shows you something remarkable: In the 1,000-year history of the Earth, there is an enormous increase in the global temperature compared to any fluctuations we've seen over that 1,000 years. That's something to think about and recognize; that if we're to prevent that, or at least moderate that, we've got a lot to do and we've got to start doing it now.

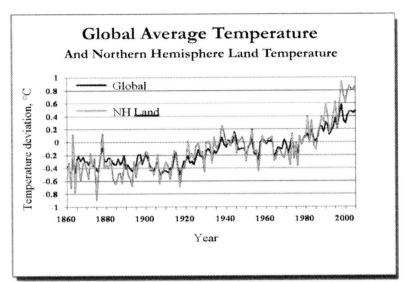

Figure 1

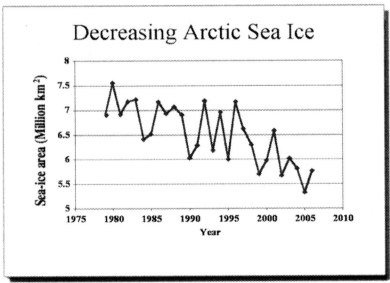

Figure 2

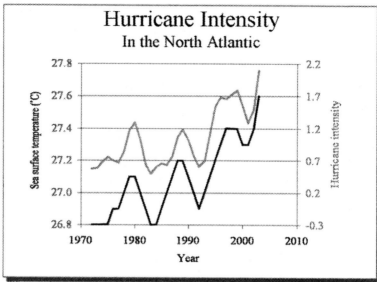

Figure 3

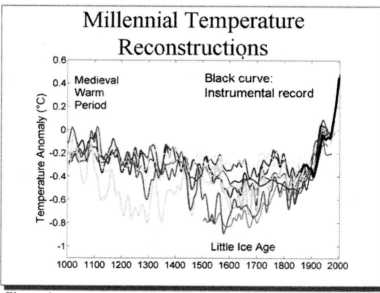

Figure 4

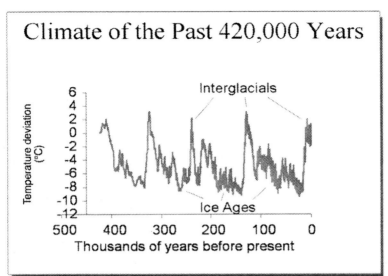

Figure 5

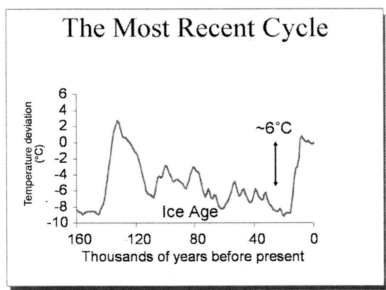

Figure 6

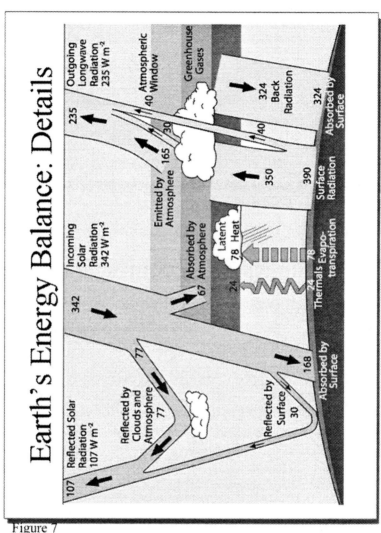

Earth's Energy Balance: Details

Figure 7

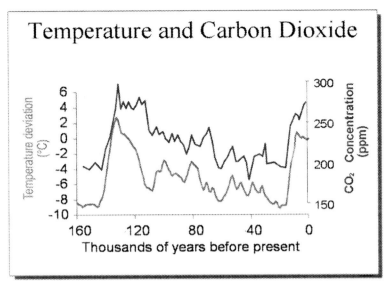

Figure 8

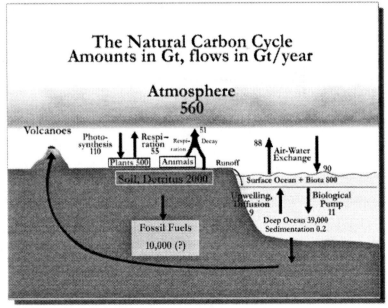

Figure 9

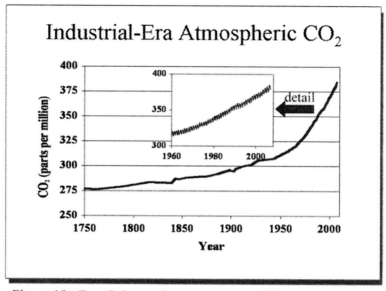

Figure 10   Detail shows the very accurate CO_2
measurements made since the 1950s.

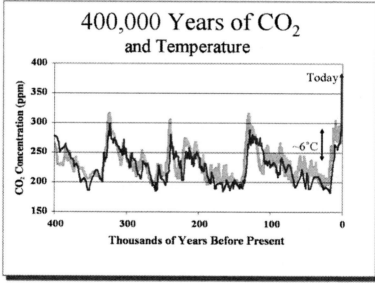

Figure 11

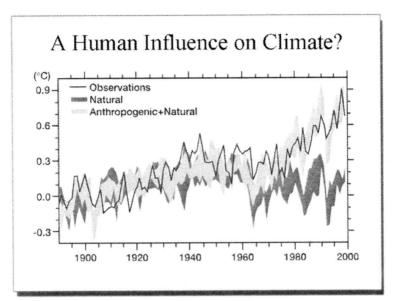

Figure 12

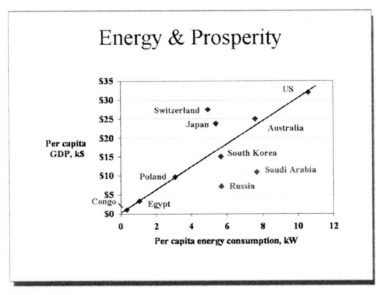

Figure 13

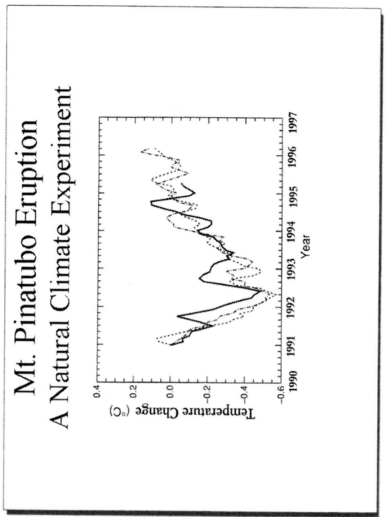

Figure 14

CO₂ and Temperature Projections

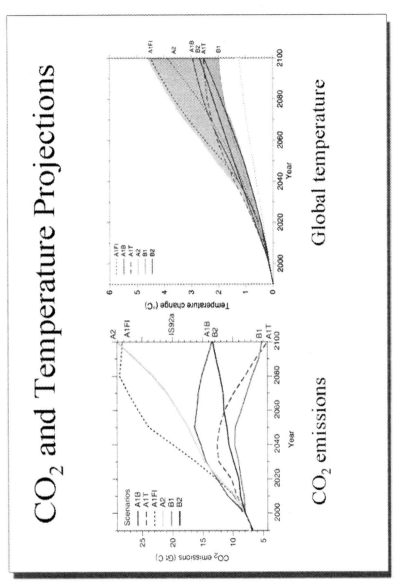

Figure 15

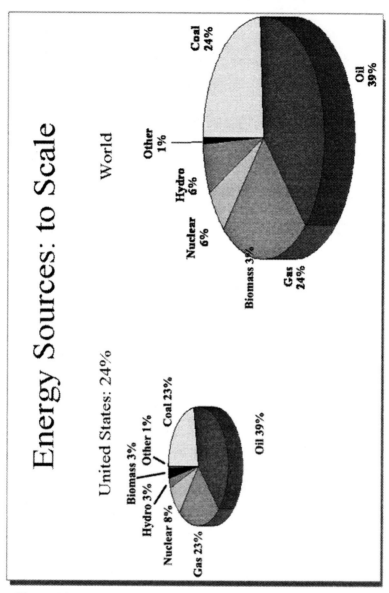

Energy Sources: to Scale

World

United States: 24%

Figure 16

Earth's Energy Resources

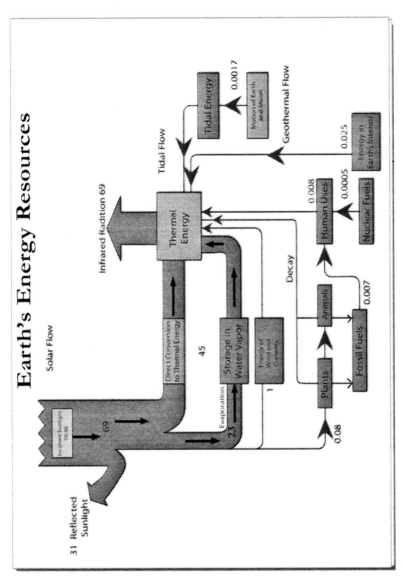

Figure 17

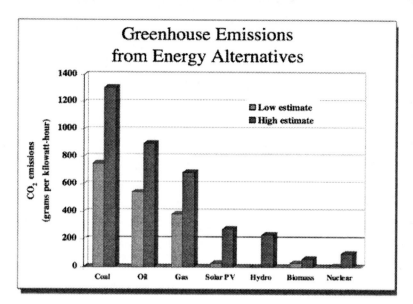

Figure 18

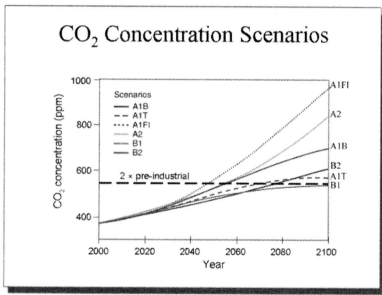

Figure 19

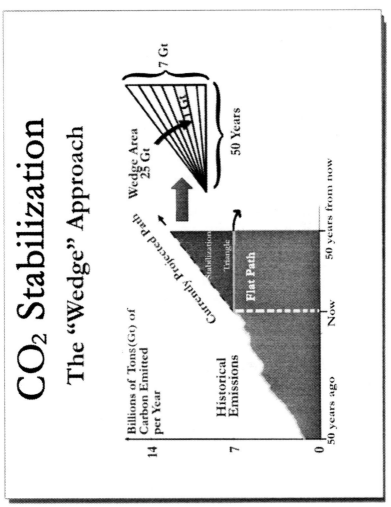

Figure 20

Data Sources for Figures

Figure 1: © Crown copyright 2007, data supplied by the Met Office Hadley Centre and CRU; Brohan, P., Kennedy, J., Harris, I., Tett, S.F.B. and Jones, P.D., 2006: Uncertainty estimates in regional and global observed temperature changes: a new dataset from 1850. *J. Geophys. Res.* 111, D12106, doi:10.1029/2005JD006548.

Figure 2: National Snow and Ice Data Center.

Figure 3: The Teaching Company. Data provided by Kerry A. Emanuel: http://wind.mit.edu/~emanuel/Papers_data_graphics.htm.

Figure 4: Robert A. Rohde / Global Warming Art.

Figures 5, 6, 8, 11: Jouzel, J., C. Lorius, J.R. Petit, C. Genthon, N.I. Barkov, V.M. Kotlyakov, and V.M. Petrov. 1987. Vostok ice core: a continuous isotope temperature record over the last climatic cycle (160,000 years). *Nature* 329:403-8; Jouzel, J., N.I. Barkov, J.M. Barnola, M. Bender, J. Chappellaz, C. Genthon, V.M. Kotlyakov, V. Lipenkov, C. Lorius, J.R. Petit, D. Raynaud, G. Raisbeck, C. Ritz, T. Sowers, M. Stievenard, F. Yiou, and P. Yiou. 1993. Extending the Vostok ice-core record of palaeoclimate to the penultimate glacial period. *Nature* 364:407-12.

Figure 7: Kiehl, J. T., and K. E. Trenberth, 1997. Earth's annual global mean energy budget. *Bulletin of the American Meteorological Society* 78, 206, figure 7.

Figure 9: The Teaching Company. Adapted with permission from Wolfson, *Energy, Environment, and Climate* (W.W. Norton & Co.).

Figure 10: Neftel, A., H. Friedli, E. Moor, H. Lötscher, H. Oeschger, U. Siegenthaler, and B. Stauffer. 1994. Historical CO_2 record from the Siple Station ice core. In *Trends: A Compendium of Data on Global Change.* Carbon Dioxide Information Analysis Center, Oak Ridge National Laboratory, U.S. Department of Energy, Oak Ridge, Tenn., U.S.A.; D.M. Etheridge, L.P. Steele, R.L. Langenfelds, R.J. Francey, J.-M. Barnola and V.I. Morgan. 1998. Historical CO_2 records from the Law Dome DE08, DE08-2, and DSS ice cores. In *Trends: A Compendium of Data on Global Change.* Carbon Dioxide Information Analysis Center, Oak Ridge National Laboratory, U.S. Department of Energy, Oak Ridge, Tenn., U.S.A.; Keeling, C.D. and T.P. Whorf. 2005. Atmospheric CO_2 records from sites in the SIO air sampling network. In *Trends: A Compendium of Data on Global Change.* Carbon Dioxide Information Analysis

Center, Oak Ridge National Laboratory, U.S. Department of Energy, Oak Ridge, Tenn., U.S.A.

Figure 12: Parallel Climate Model/DOE/UCAR; Meehl, G.A., W.M. Washington, C.M. Ammann, J.M. Arblaster, T.ML. Wigley and C. Tebaldi. 2004. Combinations of Natural and Anthropogenic Forcings in Twentieth-Century Climate. *Journal of Climate*, 17, 3723, figure 2(d).

Figure 13: The Teaching Company. Data selected from Key World Energy Statistics 2006 © OECD/IEA, 2006, pp. 48-57.

Figure 14: Intergovernmental Panel on Climate Change
IPCC, 1996: Climate Change 1995: The Science of Climate Change, p. 258.

Figure 15: Intergovernmental Panel on Climate Change
IPCC, 2001: Climate Change 2001: The Scientific Basis, p. 64.

Figure 16: Energy Information Administration
EIA, Annual Energy Review 2004, Table 1.3; EIA, International Energy Annual 2003, Table 1.8.

Figure 17: The Teaching Company. Adapted with permission from Wolfson, *Energy, Environment, and Climate* (W.W. Norton & Co.) and Romer, Robert H. 1985. Energy Facts & Figures (Spring Street Press, Amherst, MA). Romer is the original source for the figure.

Figure 18: The Teaching Company. Data from International Atomic Energy Agency, excepting the nuclear high value. Spadaro, J.V., Langlois, L., and Hamilton, B. 2000. Greenhouse Gas Emissions of Different Electricity Generating Chains, IAEA Bulletin, 42 (2). Nuclear high value is an estimate based on a survey of the literature.

Figure 19: Intergovernmental Panel on Climate Change
IPCC, 2001: Climate Change 2001: The Scientific Basis, p. 222.

Figure 20: The Teaching Company. Sources:
Marland, G., T.A. Boden, and R. J. Andres. 2006. Global, Regional, and National CO_2 Emissions. In *Trends: A Compendium of Data on Global Change*. Carbon Dioxide Information Analysis Center, Oak Ridge National Laboratory, U.S. Department of Energy, Oak Ridge, Tenn., U.S.A. ; Socolow, R. 2005. Stabilization Wedges: Mitigation Tools for the Next Half-Century. Presentation at Avoiding Dangerous Climate Change: A Scientific Symposium on Stabilisation of Greenhouse Gases. February 3, 2005. Met Office, Exeter, UK. The presentation was based on Pacala, S. and R. Socolow. 2004. Stabilization Wedges: Solving the Climate Problem

for the Next 50 Years with Current Technologies, Science, 305, 968-972.

Glossary

active solar: Technologies that use mechanical or electrical components to facilitate energy collection and storage.

aerosols: Particulate matter suspended in the atmosphere.

albedo: The reflectivity of a surface or material, expressed as a number between 0 and 1 that gives the fraction of incident light that is reflected.

anthropogenic: Of human origin.

biological pump: The process whereby dead marine organisms and their waste products sink into the deep ocean, taking carbon with them.

carbon capture and sequestration: The process of capturing carbon dioxide either before or after combustion and storing it where it cannot get into the atmosphere.

carbon cycle: The cycling of carbon through the atmosphere, biosphere, oceans, and Earth's crust.

CCS: See **carbon capture and sequestration**.

cell: The smallest division of Earth's surface, atmosphere, or ocean in a climate model.

climate: The long-term average patterns of temperature, precipitation, humidity, and other weather-related quantities.

climate model: A mathematical representation of the climate system, usually implemented in a computer program; used to study climate processes and to explore projections of future climate.

coupled model: A climate model that couples together two or more separate models, typically of the surface/atmosphere and, separately, the oceans.

cryosphere: Earth's frozen sector, comprising the polar ice caps, floating sea ice, and glaciers.

diurnal temperature range: The typical variation in temperature from night to day at a given location.

energy balance: The condition in which a system loses as much energy as it gains, thus maintaining a constant temperature.

energy intensity: A measure of the energy required to produce a given unit of gross domestic product.

enhanced greenhouse effect: The trapping of additional infrared, with subsequent warming of Earth's surface, by greenhouse gases of anthropogenic origin.

equilibrium: A state of balance in which nothing changes.

feedback effect: A change in a system that occurs as a result of another change, either enhancing (positive feedback) or diminishing (negative feedback) the original change.

fission: The splitting of heavy nuclei to release energy.

flow: A stream of energy, such as sunlight coming to Earth.

fossil fuels: The fuels coal, oil, and natural gas, formed over millions of years from organic material trapped underground before it has a chance to decay fully. The energy contained in fossil fuels originated in sunlight and was captured through photosynthesis.

fuel: A material substance that stores energy.

fusion: The joining of light nuclei to release energy.

GCM: Global climate model or global circulation model.

gigatonne: One billion tonnes, a unit useful in describing global carbon flows.

greenhouse effect: Warming of a planet's surface caused by the absorption and re-radiating of outgoing infrared radiation by gases, clouds, or aerosols.

greenhouse gas: A gas that absorbs infrared radiation, thus contributing to the greenhouse effect.

hydrologic cycle: See **water cycle**.

ice age: A long period of cool climate, characterized by mile-thick ice sheets covering the polar regions and extending, in the northern hemisphere, into what is now the northern United States.

indirect aerosol effect: A secondary effect whereby aerosols act as cloud condensation nuclei, with the resulting clouds then reflecting sunlight to produce a cooling effect.

indirect solar: Energy technologies based on wind, water, or biomass. The ultimate source of their energy is sunlight, but they do not use sunlight directly.

interglacials: Relatively brief periods of warmer conditions occurring between the much longer ice ages.

isotopes: Versions of the same element that differ in the number of neutrons in their nuclei and, thus, in mass. Subtle changes in the concentration of different isotopes provide proxies for temperature and other quantities in the distant past.

latent heat: Energy stored by virtue of a substance's being in a gaseous or liquid phase. The energy can be released as sensible heat when the gas condenses to liquid or the liquid solidifies.

MAT: Marine air temperature.

metric ton: 1000 kilograms, or 1.1 U.S. tons. Also, tonne (t).

model validation: Any procedure used to demonstrate that a climate model correctly reproduces climatic conditions.

Moore's law: The exponential increase in computing power that has computer speeds doubling roughly every 18 months.

natural greenhouse effect: The natural warming of Earth's surface through the greenhouse effect resulting from naturally occurring water vapor and carbon dioxide. The natural greenhouse effect keeps Earth's surface 33°C (about 60°F) warmer than it would otherwise be.

negative feedback: A feedback effect that diminishes the original change.

nuclear fuel: A fuel whose energy is stored in its atomic nuclei.

passive solar: Technologies that harness solar energy without moving parts, such as pumps, fans, and the like.

photosynthesis: The process whereby green plants take in carbon dioxide and sunlight to manufacture sugars, which store energy that ultimately comes from the Sun.

photovoltaic cells: Devices that convert sunlight directly into electricity.

positive feedback: A feedback effect that enhances the original change.

power: The rate of energy use or production.

proxy: A measurable quantity whose value helps determine the value of a quantity that can't be measured directly—as in the use of tree-ring data to calculate temperatures when thermometric records aren't available.

reservoir: A system that acts as temporary or long-term storage for materials cycling in the Earth system. Living things, for example, constitute a reservoir for carbon.

resolution: The fineness of the scale into which a climate model divides Earth's surface, atmosphere, and ocean. Higher resolution means a finer scale.

respiration: The opposite of photosynthesis; living things take in oxygen, which reacts with food to release energy and carbon dioxide.

SAT: Surface air temperature.

SST: Sea-surface temperature.

stratosphere: An atmospheric layer extending from the top of the troposphere to about 50 kilometers, in which most solar ultraviolet radiation is absorbed.

sub-grid parameterization: A procedure used to represent phenomena, such as the structure of clouds, that are smaller than the cell size of a climate model.

sulfate aerosols: Aerosol particles formed from sulfate (SO_4) compounds. Sulfate aerosols are highly reflective and, thus, produce a cooling effect.

thermohaline circulation: Oceanic circulation driven by a combination of temperature and salinity differences.

tonne: See **metric ton**.

troposphere: The lower level of the atmosphere, extending to about 8 to 18 kilometers, in which nearly all weather phenomena occur.

urban heat island effect: An apparent warming due to the growth of cities, which are warmer than their rural surroundings.

water cycle: The cycling of water through the Earth system, by evaporation, precipitation, and the flow of rivers and groundwater.

watt: A unit of power—the rate of energy use or production; 1 watt is 1 joule per second and is about 1/100 of the power output of a typical human body.

weather: The particular conditions that describe the state of the atmosphere in a given location and at a given time.

Bibliography

Readings:

Glantz, Michael. *Climate Affairs: A Primer*. Washington, DC: Island Press, 2003. A general-audience discussion of climate change and its impacts by a social scientist who is the former head of the Environmental and Societal Impacts Group at the National Center for Atmospheric Research.

Harvey, L. D. D. *Global Warming: The Hard Science*. Essex, England: Prentice Hall, 2000. A thorough exploration of the science behind global climate change. Doesn't spare the math but relegates the most complex math to boxes that the reader is free to skip.

Houghton, John. *Global Warming: The Complete Briefing*, 3rd ed. Cambridge: Cambridge University Press, 2004. A thoroughly readable, up-to-date account of climate change by a former chair of the Scientific Assessment Working Group of the Intergovernmental Panel on Climate Change.

Intergovernmental Panel on Climate Change. *Carbon Capture and Storage* (IPCC 2005). Cambridge: Cambridge University Press, 2005. A detailed report suggesting that CCS could play a significant short-term role in mitigating the effects of fossil-fuel combustion on climate. Also available at http://www.ipcc.ch.

———. *Climate Change 2001: The Scientific Basis* (IPCC 3). Cambridge: Cambridge University Press, 2001. This is the science portion of the IPCC's third assessment report—a comprehensive survey of the state of climate science at the turn of the 21st century. Superceded by IPCC 4 but still full of useful background information. Also available at http://www.ipcc.ch.

———. *Climate Change 2007: The Physical Science Basis* (IPCC 4). Cambridge: Cambridge University Press, 2007. This is the science portion of the IPCC's fourth assessment report—a comprehensive survey of the state of climate science in the early 21st century. Also available at http://www.ipcc.ch.

Lovins, Amory B., E. K. Datta, O-E Bustnes, J. G. Koomey, and N. J. Glasgow. *Winning the Oil Endgame: Innovation for Profits, Jobs, and Security*. Snowmass, CO: Rocky Mountain Institute, 2005. A quantitative and elaborately documented report showing how humanity might end its dependence on climate-changing fossil fuels. Also available in full at http://www.oilendgame.com.

Parmesan, Camille, and Gary Yohe. "A Globally Coherent Fingerprint of Climate Change Impacts Across Natural Systems." *Nature* 421 (2 January 2003), pp. 37–42. This paper presents a "metastudy" of hundreds of different research results that point to climate-change impacts on species ranges.

Ruddiman, William. *Earth's Climate: Past and Future*. New York: W.H. Freeman, 2001. An authoritative and beautifully illustrated textbook on climate, with emphasis on paleoclimate.

Schneider, Stephen, Armin Rosencranz, and John O. Niles, eds. *Climate Change Policy: A Survey*. Washington, DC: Island Press, 2002. This is a collection of essays dealing mostly with policy issues related to climate change. But the first chapter is a concise introduction to the science of climate change.

Smil, Vaclav. *Energy: A Beginner's Guide*. Oxford, England: Oneworld Publications, 2006. A quick guide to energy by a leading specialist and prolific writer on energy issues. Particularly strong on energy history.

———. *Energy at the Crossroads: Global Perspectives and Uncertainty*. Cambridge, MA: MIT Press, 2003. A thorough, scholarly, but eminently readable account of energy history, alternatives, and possible energy futures from a truly independent thinker on the subject.

Pacala, Stephen, and Robert Socolow. "Stabilization Wedges: Solving the Climate Problem for the Next 50 Years with Current Technologies." *Science* 305 (13 August 2004), pp. 968–972. A visionary paper that shows how we might break the seemingly insurmountable climate problem into manageable pieces.

Stern, Nicholas. *The Economics of Climate Change: The Stern Review*. Cambridge: Cambridge University Press, 2006. A thorough review of the potential economic impacts of climate change by a leading British economist and former chief economist at the World Bank.

Stott, Peter A., Dáithí A. Stone, and Myles R. Allen. "Human contribution to the European heatwave of 2003." *Nature* 432 (December 2, 2004), pp. 610-614. This paper makes a statistical argument that the European heat wave of 2003 is unlikely to have resulted from natural factors alone, but that anthropogenic greenhouse warming likely contributed to the event.

Weart, Spencer. *The Discovery of Global Warming*. Cambridge, MA: Harvard University Press, 2003. A history of global warming from the 19th century onward, tracing how the idea of anthropogenic warming first dawned on scientists, then spread into the broader society.

Wolfson, Richard. *Energy, Environment, and Climate*. New York: W.W. Norton, 2007. A textbook intended for mid-level undergraduate courses that explores the science of, and connections between, energy and climate. Some math but no calculus.

Internet Resources:

British Treasury Department, http://www.hm-treasury.gov.uk/independent_reviews/stern_review_economics_clim ate_change/sternreview_index.cfm. From this site, you can download all or part of the 700-page Stern Review on the Economics of Climate Change.

Carbon Dioxide Information Analysis Center, http://cdiac.ornl.gov/. This Web site offers a host of data and graphics on global temperatures, greenhouse gases, and other climate-related parameters. Here, you can also find graphs and data on global and regional temperatures and atmospheric carbon dioxide from sites around the world on time scales ranging from decades to hundreds of thousands of years.

Climate Prediction Experiment, http://www.climateprediction.net. Participate in a worldwide climate modeling experiment! ClimatePrediction.net harnesses the spare computing power of personal computers worldwide to run multiple simulation experiments using the world's most sophisticated climate models. You can sign your own computer up to participate and keep yourself posted on the results of the ClimatePrediction experiments.

Climatic Research Unit, University of East Anglia, http://www.cru.uea.ac.uk/. Probably the most authoritative source for the instrumental temperature record discussed in Lecture One.

Hadley Centre of the British Meteorological Office, http://www.metoffice.gov.uk. The highly respected British Met (for Meteorological) Office provides a thorough introduction to the greenhouse effect and a thorough discussion of the carbon cycle through its Hadley Centre for Climate Change.

Intelligent Energy Europe, http://ec.europa.eu/energy/index_en.html. This European Commission site features myriad links to programs and information related to the European Union's vigorous pursuit of non-fossil energy.

Intergovernmental Panel on Climate Change, http://www.ipcc.ch. This site features links to the latest IPCC Assessment Report, as well as many additional reports on more specialized climate topics.

International Energy Agency (IEA), http://www.iea.org/Textbase/stats/index.asp. This site lets you generate energy data for your choice of country, region, and energy type from the IEA's vast database. A click on the Key Statistics link gets you to a free download of the annual *Key World Energy Statistics*.

National Renewable Energy Laboratory, http://www.nrel.gov/. At this U.S. Department of Energy site, you can find out about research into alternative energy technologies.

National Snow and Ice Data Center, http://nsidc.org/. Images and data showing changing conditions in Earth's cryosphere.

Pew Center on Global Climate Change, http://www.pewclimate.org/. Covers all aspects of climate change and is especially strong on helping the business community find solutions to greenhouse emissions. On the "Global Warming in Depth" page (http://www.pewclimate.org/global-warming-in-depth), the Economics and Environmental Impacts links carry more detailed information relevant to Lecture Nine.

Princeton Environmental Institute Carbon Mitigation Initiative, http://www.princeton.edu/~cmi/resources/stabwedge.htm. This Princeton University site provides a detailed description of the "wedge" approach to greenhouse stabilization, including a movie, a simulation game, and links to the original papers.

RealClimate, http://www.realclimate.org. This blog, maintained by scientists active in climate research, has technically accurate, detailed discussions of climate issues and, particularly, the attribution of climate change to human activities. Search the site on "attribution" for discussions relevant to Lecture Seven.

U.S. Department of Energy, Energy Information Administration, *Annual Energy Review*, http://www.eia.doe.gov/emeu/aer/. Published yearly and made available on the Web, the *Annual Energy Review* is a comprehensive compilation of energy statistics for the United States. There's a less detailed section entitled "International Energy." Most of the data are displayed in both tables and graphs, and the tables are available for download as spreadsheets.

U.S. Department of Energy, Energy Information Administration, *International Energy Annual*, http://www.eia.doe.gov/iea/. A more comprehensive compilation of international energy statistics.

U.S. Environmental Protection Agency (EPA), "Greenhouse Gas Emissions."
http://www.epa.gov/climatechange/emissions/index.html. This EPA site is an authoritative source for detailed statistics on U.S. greenhouse gas emissions.

World Glacier Monitoring Service, http://www.geo.unizh.ch/wgms/. This Swiss site carries yearly data on hundreds of glaciers worldwide.

Notes

Notes